Universal Reality 2.0

Clarifying the New
Theory of Everything

Edgar L. Owen

First Edition, Version 1.1, July 1, 2017

Library of Congress Cataloging-in-Publication Data

Owen, Edgar L.
Universal Reality Revealed / Edgar L. Owen – first ed.
p. cm.
Includes biographical references.

ISBN-13: 978-1548554637 (Edgar L. Owen)

ISBN-10: 1548554634 (Pbk.)

EdgarLOwen.info

CreateSpace Independent Publishing Platform

Printed in the United States of America

To my secret muse

PREFACE

This book is a concise introduction to the essentials of the new theory of Universal Reality. It's an entirely rewritten clarification of the original 476-page book and includes a number of important new developments in the theory as well. As such it's the best current introduction to the entire theory of Universal Reality, the most convincing Theory of Everything currently available.

Universal Reality presents a revolutionary new view of the fundamental nature of reality. It's a non-mathematical but scientifically consistent presentation of the little known fundamental principles underling reality in the context of a computational universe. The theory is both internally consistent and consistent with almost all the vast body of accepted science. However it revolutionizes our current out of date *understanding* of this science and explains it much more simply and elegantly and that leads immediately to all sorts of amazing new insights.

Universal Reality is simple, elegant and easy for anyone with a basic science background to understand. Anyone interested in the fundamental nature of reality, mind, time, consciousness, existence and the universe should find this book compelling, entertaining, and eye opening. It convincing clarifies the nature of reality from an entirely new perspective on the universe found nowhere else and in the process reveals an entirely new understanding of our own true natures as well.

This book presents an easy to understand comprehensive summary of the essential elements of reality and the universe that anyone can quickly grasp in just an hour or two. It assumes a general knowledge of modern physics, cosmology and computer science, at least at the popular level, and some familiarity with the great perennial issues of philosophy will be helpful, but all that's really required is the willingness to objectively explore the deepest mysteries of reality with an open mind and entirely new eyes.

This book was written primarily in an effort to further clarify and develop my own understanding of reality, but hopefully its publication will make it accessible to others as well and generate intelligent criticisms and suggestions for improvement. I personally believe it's the best, most accurate theory of reality that has so far been discovered, but reality itself is always full of mysteries and surprises and is always the final arbiter of truth.

To the extent this book is an accurate description of reality it's not something I've created, rather it's reality itself revealing itself to someone who has hopefully been able to observe and study it without projecting too much of his own personal programming and prejudices onto it. Reality is continuously revealing itself to all of us in all its awesome glory, and I believe anyone willing to observe it carefully and objectively will be able to personally verify and experience the truth of most of what this book contains.

I would like to thank everyone who has helped make this book possible and encouraged me while writing it. Thanks to all of you for putting up with my unusual hermetic life style. And a special thank you to all my wild visitors, including the occasional human, and to the beauty and profundity of nature, which always inspires me with meaning and joy. Thanks to reality itself for continuously revealing itself in all its glory to those who will only look with opened eyes, and thanks most of all to my secret muse. Thank you! Thank you! Thank you all!

And finally thanks to all those thinkers, scholars, scientists and visionaries throughout history without whose heroic efforts, genius and cumulative hard work this book could not have been written.

The author welcomes all questions and comments and can be contacted at Edgar@EdgarLOwen.com.

CONTENTS

INTRODUCTION

The goal of Universal Reality is to discover the complete Theory of Everything, to accurately explain the essentials of every aspect of the universe including both the universe of science and the world as we actually experience it.

As we will see the evidence the universe is computational is overwhelming and our universe is best understood as a running program continually computing its observable data. Universal Reality is the *systems design* for the running program of the universe. It carefully examines what kind of computational system could best account for all aspects of the universe in the simplest, most accurate and complete manner possible. To the extent Universal Reality accomplishes this it will be the systems design of the actual universe, it will be the correct Theory of Everything.

If the universe is a computational system then the best Theory of Everything will also be a computational system. Thus the idea behind Universal Reality is to design a computational system that does exactly what the actual universe does, to completely and accurately reproduce the workings of the actual universe insofar as possible in print.

Equally important is to understand how a computational system consisting of programs and data can manifest as a real actual universe. In other words what does it take to make a computational universe into a real and actual universe exactly like ours? The answer to this question reveals the nature of existence and consciousness and is one of the most important insights of Universal Reality.

A programmatic version of Universal Reality is under consideration and this book is the systems design for that program. If we wanted to construct a universe that would exactly duplicate the real universe Universal Reality should show us exactly how to do that. This is the ultimate test of the theory.

This is the goal of Universal Reality and after a lifetime of careful study Universal Reality appears to be the most comprehensive and convincing current Theory of Everything available, though of course the reader, and more importantly the universe itself, will make their own judgments. It's simple, elegant, and consistent with both established science and our fundamental experience of reality as no other theory is,

1

but it also reveals a completely new and revolutionary understanding of the fundamental nature and unity of all aspects of the entire universe.

Universal Reality is a unified Theory of Everything that seamlessly integrates all aspects of reality, including time, relativity, quantum theory, cosmology, information, emergence, the nature of the present moment, our mind's simulation of reality, consciousness, and the fundamental nature of existence while maintaining complete consistency with contemporary science, logic and the scientific method with a few suggested improvements.

By discovering the simple hidden principles that underlie and unify relativity, quantum theory, and cosmology Universal Reality reveals a revolutionary new understanding of spacetime, mass-energy, and the computational nature of reality that leads to all sorts of important new insights.

Universal Reality Revealed takes us on a guided tour through this wonderful revolutionary new universe exploring and explaining every aspect of its structure based on the hidden principles that compute it. It's an amazing unified vision of the universe and reality you won't find anywhere else. And once revealed it leads naturally to a life changing experience of the deeper nature of reality in which neither the universe or our own inner selves will ever be the same.

THE COMPUTATIONAL UNIVERSE

OVERVIEW

The universe is most simply and accurately explained as a single computational system that includes everything that exists. In this view it consists entirely of data that's continually recomputed in every tick of the processor that computes it. In other words the universe is a running program that continually recomputes its current data state according to logico-mathematical rules called the laws of nature. Though the universe actually consists entirely of data at its most fundamental level our minds simulate it to us as the familiar physical universe we observe around us.

The core insight of the theory of Universal Reality is that the universe must be a computational system for it to work as it does. All this means is that at the most elemental level everything happens according to a set of consistent logico-mathematical rules called the laws of nature. In fact it's quite clear that the only way anything can happen is for it to be computed, either by a computer program, the mind of an organism, or by the universe itself. That is for anything at all to happen in a consistent organized fashion it must be the result of a consistent dynamic process that computes current states from prior states on the basis of coherent rules.

Now the only way the universe can be computational is for it to consist of data at its most fundamental level because only data can be computed. It's important to understand this doesn't mean the universe or anything in it changes. After all we, and the entire universe, are actually composed entirely of elementary particles but when we realize that everything remains exactly the same, only our understanding changes. So if those elementary particles are actually data structures everything still remains just as it was and we have the basis for a computational universe in which subsequent states can be consistently computed from previous states.

Thus we must no longer think of the universe in terms of causation. Causation must be replaced with computation. Everything that happens down to the finest detail is ultimately data being computed. That's the only way anything can happen because it's the only way all the precise fine details of things and events can be consistently transformed

3

from one state to another by logico-mathematical operations to produce the universe we actually observe.

This immediately leads to two profound insights. First that the universe is not physical or material in the traditional sense, and second that the universe itself must be a running program continually recomputing its data state. These insights free us to discover an entirely new model of the universe and the underlying reality that supports it, and this model can also seamlessly incorporate the nature of existence, consciousness, and the present moment, subjects about which contemporary science has had nothing meaningful to say.

Thus the most reasonable model of the universe is a running program that continually recomputes its data state on the basis of laws of nature that are embedded in its code. Thus there is the observable universe that consists of the current data state of all its particles, and a hidden computational process that continually recomputes it in terms of its particle interactions. From this elegant and relatively simple system the entire universe including ourselves emerges in accordance with the established laws of science.

EVIDENCE THE UNIVERSE ISN'T PHYSICAL

We think of the universe as being a physical entity but there is overwhelming evidence this simply isn't true. First it's important to understand what 'physical' actually means. Physicality is an ancient concept dating back at least to the Greeks. It's derived from the apparent solidity of matter, the qualities of its hardness, weight, and impenetrability, and of course the very name of 'physics' is rooted in this ancient concept.

But modern science reveals matter isn't solid at all and those defining qualities are due to invisible forces holding its particles together in an almost entirely empty space. So modern physics has already abandoned the concept of physicality in all but name but remains unable to take this to its logical conclusion and abandon it altogether. Physics still considers elementary particles and spacetime as the last remaining bastions of physicality but we will soon see these are best understood as computational structures as well. Thus the whole idea of physicality falls apart at the scientific level and must be replaced with the more logical and much richer concept of a computational universe consisting only of programs and data.

The non-physicality of reality is also confirmed by common sense and biology:

1. We wouldn't notice any difference at all if the universe weren't physical but consisted of programs computing data instead. We already know that everything in the universe including ourselves consists only of elementary particles. Now just imagine those elementary particles are data entities instead of physical entities. Nothing changes; we just have a new insight into the fundamental nature of everything including ourselves.
2. We already know that the universe *as we experience it* actually exists as neural data in our brains. Yet it still seems completely real and physical. So if our experience of the universe consists only of data and still seems so amazingly real then it's reasonable that the actual external reality upon which our experience is based could also consist entirely of data and be completely real as well. And how else could it be so convincingly simulated as data in our brains if it too didn't consist entirely of data as well?
3. Science, our most accurate model of reality, is entirely a logico-mathematical structure. How could science so accurately model the actual universe if the actual universe wasn't also a logico-mathematical structure? This automatically explains Wigner's "Unreasonable Effectiveness of Mathematics in the Natural Sciences" (Wigner, 1960).
4. This also immediately solves the problem of the status of the laws of nature. In a computational universe the laws of nature are simply the programs that compute the data state of the universe and are as real as the data they compute. But in the traditional entirely physical universe the laws of nature somehow stand outside of the physical universe they mysteriously control in some supernatural way in some sort of imaginary metaphysical realm. There is no place for them in the actual physical universe. This is clearly an unsatisfactory model that some of the best minds of the physical interpretation have unsuccessfully struggled with (Penrose, 2005).
5. With a little practice it's also quite easy to observationally confirm that the apparent physicality of things breaks apart into specific types of perceptual information that in combination our simulation tells us makes a physical thing. This is explained in detail in the chapter on *The Simulation*.

6. A computational model immediately opens the gates to a unified synthesis in which all aspects of reality including existence and consciousness take their places in a completely new and consistent Theory of Everything.

COMPUTING REALITY

1. At its fundamental level the universe must be a computational system for it to work as it does. That's because every precise detail of current states is always the result of logically consistent operations on previous states and the only way this can happen is for them to be computed.
2. It's impossible to imagine any other actual mechanism to produce reliable sequences of events. Modern science cops out on this just describing what sequences of events occur without offering any actual mechanism for what produces those sequences. The mechanism stated is vaguely labeled 'causation' but there is no explanation of what the actual mechanism of causation is. The only possible known mechanism that produces reliable consistent sequences of states is computation.
3. The essence of science is its discovery that repeatable logico-mathematical sequences *connect* events. But so far science has been blind to the simple obvious conclusion that the only way this can be true is for those logico-mathematical sequences to actually compute events. Thus the logico-mathematical sequences that connect events are not just *descriptions* of events but the *actual mechanism* that produces them.
4. If the universe is a computational system it must consist of its data, a running program encoding the laws of nature that computes that data, and a processor that executes the program.
5. For the universe to be computed it must consist of data at its most fundamental level because only data can be computed. And for the universe to compute its data it must be a running program because only programs can compute data.
6. Thus the universe can be envisioned as a single dynamic computational system that includes everything that exists. Nothing else exists and there is no outside, nor before or after because it includes the computation of both time and space.
7. And because the all the data of the observable universe exists directly in the medium of reality itself called the quantum vacuum rather than the medium of a printed page or a computer's memory

it becomes the data of all the real actual things of the world rather than just words on a page or a program running in a computer.

8. Thus everything is the complete information of what it is. All the things of the world are the complete information structures of what they are in the underlying medium of reality, whether they are the information of actual things or the information of descriptions of things in a book on a desk.

THE QUANTUM VACUUM

1. Physics has discovered that the quantum vacuum is the universal substrate in which virtual particles exist and from which they actualize. Thus the quantum vacuum must contain the information of the particles that can actualize from it. Thus it's reasonable to assume the quantum vacuum also supports the continuing existence of all the actualized particles that make up the observable universe. Particles actualize within it and can disappear back into it, thus their continuing existence can only be maintained by their continuing existence within it.

2. Thus everything including all actualized particles and the entire observable universe can be said to exist as data structures *within* the quantum vacuum. The quantum vacuum is the fundamental substrate and medium of the entire universe that contains and supports all its data.

3. If everything in the entire universe exists within the quantum vacuum then the quantum vacuum is the realm of existence within which every aspect of reality exists and gains its individual existence from.

4. Thus all that exists is the quantum vacuum and the virtual and actualized data that exists within it. The quantum vacuum and all the data that exists within it is the entire universe.

5. For this to be true the quantum vacuum must contain both the data of actualized particles and the virtual data that defines particles, their structures, and how they interact. It must contain the program code that computes the actualization and interactions of particles.

6. Though a single entity, the quantum vacuum can be described in terms of two coterminous realms, *the virtual universe* and the *observable universe*. The data of both realms are equally real components of the universe. Both the virtual and observable data

of the universe exist within the universal medium of existence, the quantum vacuum.

7. This immediately solves the vexing traditional problem of how non-physical laws of nature could somehow control a physical universe from a mysterious external metaphysical realm (Penrose, 2005). The actual universe is not physical in the traditional sense but composed of data just as the laws of nature are. Both exist together in the quantum vacuum as different categories of data.

8. At this fundamental level the universe is a non-dimensional *computational space*. Its elemental program and data define a computational space in the same sense that a computer program defines a non-physical, non-dimensional computational space.

9. What we think of as dimensional spacetime is computed along with everything else. Dimensional spacetime is the overall logical consistency of the dimensional relationships created among particles as their interactions are computed as explained in the chapter on *Quantum Reality*.

THE OBSERVABLE UNIVERSE

1. The observable universe is what is mistakenly called the 'physical' universe. The observable universe consists of the data of actualized elementary particles and their computational interactions and nothing else. It consists of actualized particle data continually recalculated on the basis of an elemental program running in the virtual realm of the quantum vacuum.

2. The observable universe is observable in the sense that its data forms computationally interact and thus are subject to observation. Every interaction can be considered a *mutual generic observation* in which each data form is modified by its interaction with the other. For example we observe things through the changes our interactions with them produce in our perceptual data

3. The relatively simple interactions of elementary particles produce the entire observable universe in aggregate. Thus only particle interactions need to be computed to produce the entire observable universe.

4. All the compound objects in the universe are *emergent manifestations* of elementary particle aggregates. Emergent objects and processes include all the things we observe in the world around us at the classical scale.

5. The data of the particles and particle components are the only actually stored data of the universe. Their states are all that are directly computed by the elemental program. As explained in the chapter on *Emergence,* calculations of elementary particle interactions are all that is necessary to compute the entire observable universe because all larger scale phenomena are *emergent patterns* that automatically manifest in aggregates of particle interactions due to the specifics of the complete fine-tuning.
6. Thus almost all of the equations of science are simply *descriptions* of particle behavior at the aggregate scale and aren't part of the *computations* that directly compute them. This vastly simplifies the computational system of the observable universe and the Theory of Universal Reality that describes it.

THE VIRTUAL REALM

1. Science already recognizes a realm of virtual data called the quantum vacuum. So it makes sense to assume all the virtual data necessary to compute the entire observable universe exists in the virtual realm of the quantum vacuum. In this view science has just begun to discover the entire realm of virtual data in the quantum vacuum. Where else could the unobservable data necessary to compute the observable universe exist?
2. Virtual data means the data is observable only by its effects on the observable data. Both virtual and observable (actualized) data are equally real components of the universe. All the data necessary to compute the entire observable universe must exist as virtual data in the virtual realm of the quantum vacuum.
3. The total set of specific values and forms of all the virtual data necessary to compute the observable universe can be called the *complete fine-tuning.*
4. The complete fine-tuning includes the data of the basic rules of logic, data manipulation, and relatively simple mathematics embodied in *the fundamental operators* (the 'machine language' of the universe). It also includes the *fundamental laws of nature*, the *fundamental constants*, the *structural templates* that determine the elemental structures of particles and spacetime. And it includes a single *elemental program* that uses this virtual data to compute all particle evolutions and interactions. The virtual data

contains everything necessary to compute every aspect of the observable universe in terms of its particle interactions.

5. The virtual realm also contains the templates for the particle components that combine to form the real particles of the observable universe. It can be considered as a reservoir of unactualized particle components.

6. All this virtual data is intrinsic to the quantum vacuum and universally available to all the computations of the observable universe.

THE UNIVERSAL PROCESSOR

1. Now if the universe is computational then there must be a processor that computes it. Like a computer the universe must have a processor that drives its computations by executing the code of the elemental program to calculate particle interactions.

2. This processor is an intrinsic attribute of the quantum vacuum essential to its existence. It's the source of what can be called 'happening'. *Happening* manifests as computational change and is the ultimate source of the flow of time. It's the ultimate source of the life and energy of the universe.

3. Because it exists the universe must have presence. The presence of the universe manifests as a *universal current present moment* common to the entire universe in which all its particle data is recomputed including all its different clock time rates.

4. Contrary to common belief, a universal present moment is entirely consistent with relativity in a computational universe when all local relativistic clock time rates are simultaneously computed by the processor in a universal present moment.

5. Thus there are two distinct kinds of time, present moment time which can be called *P-time*, and clock time, which runs at different relativistic rates within P-time. This is examined in detail in the chapter on *Time*.

6. Thus the entire universe exists in the single current universal present moment it manifests by its presence and happening. Everything that exists does so within this single universal present moment including all clock times running at local relativistic rates. This is what we personally experience as the present moment in which we all exist and through which our clock times can flow at different rates.

7. The processor executes all the computations of the universe simultaneously in the present moment because happening is an intrinsic attribute of the quantum vacuum within which all the data of the universe exists.
8. Each universal simultaneous recomputation of the entire data state of the observable universe creates a single tick of universal P-time. It generates the current present moment time and data state of the entire observable universe. The concept of a universal present moment time is consistent with the disparate clock rates of relativity because relativistic clock time is simultaneously computed for all processes based on the local amount of spatial velocity as explained in the upcoming chapter on *Relativity & Spacetime*.
9. A separate *application* of the processor and the elemental program are used to compute each coherent process independently as explained in the chapter on *Quantum Reality*. This enables us to explain quantum entanglement.

RELATIVITY & SPACETIME

ENERGY AND MASS

A fundamental insight of Universal Reality is that all forms of energy including mass are simply explained as different forms of relative spatial velocity. We already know that kinetic and heat energy are both energies of linear spatial velocity, and that electromagnetic energy is the wave frequency energy of photons. It's a simple natural extension to model mass as the energy of fields of very fine intrinsic vibrational velocity in space, and electromagnetic fields as clockwise and counterclockwise helical rotations in space.

This principle immediately provides the previously unexplained reason underlying the conservation of mass and energy. The conservation of energy is now simply a matter of transforming equivalent amounts of one form of spatial velocity to another. It is simple logic that for different things to be conserved they must actually be different interconvertible forms of the same thing.

This insight also immediately explains the true nature of gravitation and its relativistic effects, and demonstrates how gravitational time dilation and linear time dilation are two aspects of the same phenomenon, the reduction in the velocity of time due to increased velocity in space. This in turn leads to an extremely simple and elegant new model of relativity, and on to an entirely new understanding of spacetime itself.

RELATIVITY MADE SIMPLE

There is no doubt at all that the *equations* of relativity are correct. They've been extensively confirmed by tens of thousands of observations for over 100 years, and are routinely used to compute the trajectories of space probes, explain the workings of the universe, and even GPS wouldn't work correctly without them.

But we must carefully distinguish between the actual equations of relativity, namely the Einstein field equation, the geodesic equation and their derivative equations like the Lorentz transforms, and the extensive *interpretive model* meant to explain these equations, which often incorrectly passes for part of the actual theory.

The equations work splendidly and their accuracy is extensively confirmed, however the interpretive model can be greatly improved to place relativity in the context of a single unified Theory of Everything that remains consistent with its equations.

For example few people realize there are two simple hidden principles that underlie relativity and make it an obvious consequence of the nature of spacetime itself and very easy to understand. And these principles immediately lead to a number of other profound insights about the universe.

The two fundamental unifying principles that explain relativity can be called:

1. **The STc (space, time, speed of light) Principle**: Everything in the universe continually moves through combined space and time at the speed of light c. This is true in all cases and for all observers when the next principle is included.
2. **The MEv (mass, energy, velocity) Principle**: All forms of mass and energy are forms of spatial velocity. In particular masses are fields of intrinsic vibrational spatial velocity centered on their sources.

These two principles can be combined into a single fundamental principle that governs the relativistic spacetime universe:

The METc (mass, energy, time, speed of light) Principle: Everything in the universe always has a combined space plus time velocity equal to c, the speed of light, and the amount of this c velocity that is spatial always manifests as some form of energy or as mass. Thus the total mass-energy plus time velocities experienced by everything in the universe are always equal to c. This is the single fundamental principle that underlies all of relativity.

[Note it's the *vector sum* of space and time velocities that is always equal to c. This means that it's actually the sum of the squares of speeds in space and in time that's equal to the speed of light squared.

$$v_x^2 + (cv_t)^2 = c^2$$

where v_x is spatial velocity, v_t is velocity through time and c is the velocity of light. The velocity through time is simply the number of seconds per second on some clock relative to a standard 1 second per second standard non-time dilated clock. Note that the velocity through time, v_t is multiplied by the speed of light c to put it into the same meters per second units as spatial velocity so it can be meaningfully compared to the velocity in space in a 4-dimensional universe.]

Universal Reality's unification of spacetime and mass-energy as spatial velocity in the METc Principle is consistent with the fact that Einstein derived his famous $E=mc^2$ relationship of mass and energy directly from the nature of spacetime in special relativity (Wikipedia, *Mass-energy equivalence*).

TIME & LINEAR VELOCITY

Everything in the universe has both a clock time rate (a velocity in time), and a velocity rate in space. By The STc Principle the vector sum of these two rates is always equal to the speed of light. Thus the more spatial velocity an object has the less time velocity it will have. In other words the presence of spatial velocity in any form always slows velocity in time. Any clock that is moving in space will be running slower.

Now a basic tenet of special relativity is that clocks appear to run at different rates depending on whether they are moving relative to an observer or not. An observer's own (comoving) clock obviously has no relative spatial motion to the observer and so by STc it always seems to run at c. Thus every observer and everything in the universe always moves at the speed of light through *time* on its own comoving clock. The velocity of time on an observer's comoving clock is called his *proper* time and is always equal to c. Thus everything in the universe, including us, is always moving through time at the speed of light on its own clock. This is what we experience as the passage of time.

But if a clock is moving relative to an observer the STc Principle dictates that clock will appear to run slower so that the vector sum of its velocity through time and its spatial velocity always remains equal to c. This slowing of time due to velocity in space is referred to as *time dilation*.

14

This STc Principle holds for all observers observing any clock with any spatial velocity in the universe. It's truly one of the most fundamental principles of the universe, and when gravitation is included it very simply explains the essence of relativity.

MASS & GRAVITATION

The next fundamental principle is what's needed to explain gravitation and its affect on time. It's fundamental to understanding general relativity.

The MEv Principle tells us that mass, being a form of energy, must actually be some form of spatial velocity. To explain the known properties of mass it needs to be modeled as a field of intrinsic spatial velocity in the form of very fine scale standing vibrations in space centered on its source. Thus a gravitational field is an actual part of its central mass rather than some mysterious thing produced by a mass. Masses actually *are* gravitational fields centered on their source particles rather than just points in space. This greatly simplifies our model of the universe because we no longer have to explain how masses actually produce gravitational fields, something missing from traditional physics.

The important insight that charges are actually fields centered on their source particles rather than points in space is a fundamental principle. Physics tends to conceive of fields as something charges produce in some mysterious manner, but it's much simpler and more reasonable to consider fields as what charges actually are, and charges as fields that extend out into space. All the charges of all the four fundamental forces can be modeled as their fields.

If we take gravitational fields as fields of intrinsic spatial velocity the relativistic effects of mass immediately become clear. These vibrational fields are fields of intrinsic spatial velocity in the fabric of space that by STc automatically reduce the time velocity of anything in the field.

Thus we automatically get gravitational time dilation and the other effects of general relativity from this very simple model of masses as fields of vibrational spatial velocity.

We also automatically get a new explanation of gravitational attraction since at every point in the intrinsic velocity field there is now a velocity vector pointing towards the source of the gravitating mass because the intrinsic velocity density is greater in that direction. Thus inertial motion tends to follow these velocity vectors towards the center of the gravitational field and we have an automatic explanation of why gravitational fields attract things.

So objects can have linear velocities in space, and they also experience the intrinsic spatial velocities of any gravitational fields because they are riding its vibrations. Thus an object's velocity in time is reduced by both its own linear velocity and by the intrinsic spatial velocities of gravitational fields.

Thus the METc principle explains both linear and gravitational time dilation as consequences of a single principle, something missing from usual interpretations of relativity. In addition the *relativistic increase of observational mass with linear velocity* is now revealed as the same as the *increase of weight* (observational mass) in a stronger gravitational field.

COMPUTING THE STc PRINCIPLE

Everything in the universe is computed by the processor of happening. One of the things the processor computes is the allocation of the c valued velocity of every process between spatial and time velocity. In doing so the processor computes the clock time rates of all processes in the universe.

Universal Reality provides a simple elegant mechanism for how the universal processor computes the dimensional spacetime aspects of both relativity and quantum reality as unified aspects of a single fundamental process.

Assume a separate *application* of the elemental program is used to simultaneously compute each separate coherent process of the universe.

Now assume each application computes the spacetime dimensionality of each new data state in terms of a *fixed number of cycles*. The total number of these cycles will be such as to always produce a total c spacetime velocity for every process it computes.

Assume the spatial velocity of each process is computed first. Then whatever number of processor cycles remains is used to compute the internal evolution of the process. The internal evolution of a process manifests as its velocity in time. So this simple allocation of a fixed number of processor cycles to computing first velocity in space and second velocity in time automatically manifests as the computational source of the STc Principle.

Thus the simple allocation of a fixed number of processor cycles to computing velocity and space and then velocity in time automatically generates the core principle of relativity.

Not only that but a simple tweak to this processor model also explains quantum wavefunctions and the Uncertainty Principle since they both reduce to a conflation of space versus time velocities as explained in the upcoming chapter on *Quantum Reality*.

AN ELEGANT NEW MODEL OF SPACETIME

This model immediately gives us a new understanding of the fabric of spacetime itself. Spacetime is now seen as a universal field of c valued velocity, which can either be velocity in space or velocity in time or any combination so long as their vector sum remains equal to the speed of light c. In this simple and elegant model relativity becomes a natural and easy to understand consequence of the nature of spacetime itself.

In empty space devoid of mass or energy that c valued velocity is entirely through time (neglecting the zero-point energy effect explained later). But the presence of the intrinsic velocity vibrations of a gravitational field reduces velocity in time within the points of the field so the vector sum of spatial and temporal velocity remains equal to c.

This means the spacetime fabric can be considered a universal energy field whose strength at any point is the amount of spatial velocity, and the remainder of its c valued velocity is expressed as velocity through time. In this model mass and electromagnetic charges are simply spherical volumes of space where some of the c valued baseline velocity of time has been converted to that of intrinsic spatial velocity. Thus these charges actually are conversions of time velocity to that of spatial velocity in forms particular to the type of charge.

Though we tend to think of spacetime in terms of expanses and positions it's fundamentally a field of fixed c valued velocity. Thus spacetime includes the fundamental concept of continuous change or happening in itself. Distance, in either time or space, is a derivative concept resulting from velocity.

Thus all spacetime is a uniform field of c valued velocity that can take the form of either the energy of velocity in space or velocity in time, or any combination so long as their vector sum is equal to the speed of light c.

When velocity is expressed as any form of velocity in space that manifests as some form of energy. This includes the energy of mass, which takes the form of fields of very fine vibrational standing waves.

The traditional definition of mass-energy is equivalent to spatial velocity of one form or another, however the velocity of time is another aspect of the universal velocity fabric and can also be considered an energetic process. Together, velocity in time and velocity in space are dual aspects of the energetic happening that brings the universe and everything in it to life. Everything in the universe is either some form of mass-energy or velocity in time or some combination.

Thus mass is a field of intrinsic spatial velocity that reduces local clock time velocities. Charges like mass aren't fields *in* space they are fields *of* space, and space itself is a single field of c valued energy.

This includes the zero-point energy of the quantum vacuum, which appears to be constant throughout space. Every point in the universal spacetime field is characterized by a very small zero-point energy that consists of the aggregate spatial velocity of virtual particles popping in and out of existence.

Both the uniform field of spacetime and all particles and particulate objects within it obey the same rule that their total spacetime velocity is always equal to the speed of light c.

In all cases both points in the field of spacetime and particles within it all have a fixed uniform space plus time velocity equal to c that takes the form of either c velocity in space or c velocity in clock time or a combination that vector sums to c. The entire observable universe is a field of c valued velocity or happening that can be expressed as any combination of velocity in space and velocity in time so long as they equal the speed of light.

But actually the value of c has nothing to do with light. It's actually the fixed velocity of spacetime and more fundamentally the number of fixed processor cycles the processor uses to compute the happening of the universe. Light just happens to have no velocity in time on its own comoving clock and thus always travels at the speed of light in space.

Even though both points in spacetime and particles at those points are both subject to the STc Principle the total velocity always remains equal to c because gravitational fields automatically reduce the total spacetime velocity of objects within them. This is explained in detail in the section on *Black Holes*.

ADVANTAGES OF THIS NEW MODEL

This new understanding of relativistic spacetime has profound consequences and greatly simplifies our view of the observable universe.

1. It replaces the usual curved spacetime model of relativity with a completely equivalent flat Euclidean fixed c velocity spacetime where gravitational fields take the form of intrinsic densities of spatial velocity. Thus it models spacetime as the flat Euclidean structure we actually observe rather than an impossible to visualize 4-dimensional curvature. This makes the universe much easier to comprehend. In this model objects curve around masses as before but because they follow gradients of vibrational density rather than following invisible curvatures of space.

2. So spacetime is now flat and Euclidean with every point having a fixed c velocity. To the extent that velocity is spatial velocity a gravitational field is present. The remainder of the fixed c velocity density manifests as the velocity through time at that point. Relativity becomes a simple natural consequence of the nature of spacetime itself.

3. The fact that the traditional curved spacetime model is not an essential part of relativity but only an *interpretation*. In general terms the Einstein field equation expresses the effects of mass-energy on spacetime as changes in the volume of points in space. These deviations in volume are normally interpreted as *dilations*

in a fixed spacetime grid that produce curvatures of the grid lines but this is equally compatible with the completely equivalent flat velocity density spacetime model of Universal Reality.

4. Both gravitational and linear velocity time dilation have the same relativistic source of time velocity being reduced in the presence of spatial velocity.

5. We also have a direct explanation of how mass affects spacetime that is missing from traditional relativity which offers no explanation for why mass 'curves' spacetime. Mass actually is a field of intrinsic spatial velocity density in the form of very fine vibrations.

6. The model simply explains the previously unknown reason for the conservation of mass and energy. The conservation of different forms of mass and energy is now simply the conversion of one form of spatial velocity to an equivalent amount of another form of spatial velocity. It's really quite obvious that only different forms of the same fundamental thing can be conserved.

7. The underlying nature of spacetime as a field of uniform c valued space plus time velocity is revealed in a manner that makes relativity a simple natural consequence of spacetime itself rather than a confusing add on.

8. It's also much easier for a computational universe to compute events in a flat Euclidean space rather than the curved spacetime model of relativity.

Thus by recognizing two hidden principles we arrive at a very simple and elegant new interpretation of relativity that leads directly to a new understanding of the fabric of spacetime as well. Universal Reality provides a truly revolutionary new conceptual model of mass, energy, spacetime and relativity that's completely consistent with the actual equations and confirmed observational behavior. Additional detail is provided in *Universal Reality* (Owen, 2016).

ELECTROMAGNETISM

The electromagnetic force is quite interesting because the relationship between its electric and magnetic components behave much as energy, space and time do. Just as energy is space in relative motion, so magnetism is electricity in relative motion. Electricity and magnetism are two orthogonal (90°) components of a single underlying entity just as space and time are, and in both cases each is transformed into the other via relative motion. Velocity in time is converted in velocity in space, and electric fields are converted to magnetic fields by relative motion.

In physics, a magnetic field is the relativistic part of an electric field, as Einstein explained in his 1905 paper on special relativity. When an electric charge is moving from the perspective of an observer, the electric field of this charge due to space contraction is no longer seen by the observer as spherically symmetric due to relativistic shortening along the axis of motion, and must be computed using the Lorentz transformations. One of the products of these transformations is the part of the electric field that only acts on moving charges which is called the magnetic field (Wikipedia, *Electromagnetism portal*).

This similarity between electricity and magnetism and space and time underlies the Kaluza-Klein Theory in which electromagnetism is modeled as a 5th compacted dimension and an electric charge is a standing velocity in that 5th dimension (Halpern, 2006). The beauty of this theory is that when this 5th dimension is added to the 4 dimensions of general relativity, Maxwell's equations of electromagnetism automatically emerge (Wikipedia, *Kaluza-Klein theory*).

Thus one can consider electricity as a fundamental force and magnetism as electric charge(s) in motion. This motion can take several forms. At the elemental level of particle components all electrically charged particles have an intrinsic half integer spin, which effectively rotates the charge about an axis.

Because spin gives its associated electric charge rotational motion spin manifests as magnetism and particle spin is the intrinsic underlying unit of magnetism. Since spin about an axis creates an axis and produces an orientation of the axis, spin is equivalently an intrinsic underlying unit of angular momentum and dimensional orientation relative to the computational background. This is critical to understanding the absolute nature of rotation, as we will see in upcoming chapters.

The quantum mechanical velocity of electrons in atoms produces the magnetism of permanent ferromagnets. Ferromagnetism is due primarily to the alignment of the spins of ionic (outermost orbital) electrons in atoms. In most materials the spins of particles are randomly aligned and tend to maintain their random alignments just as spinning gyroscopes do.

Materials made of atoms with filled electron shells have a total dipole moment of zero, because every electron's magnetic moment is cancelled by the opposite moment of the second electron in the pair. Only atoms with partially filled shells (i.e., unpaired spins) can have a net magnetic moment, so ferromagnetism only occurs in materials with partially filled shells.

These unpaired dipoles (often called "spins" even though they also generally include angular momentum) tend to align in parallel to an external magnetic field, an effect called paramagnetism. Ferromagnetism involves an additional phenomenon; the dipoles tend to align spontaneously, giving rise to a spontaneous magnetization, even in the absence of an applied field (Wikipedia, *Ferromagnetism*).

A fundamental characteristic of magnetism is because it's due to rotational spin about an axis and the poles of the spin axis are spinning in opposite directions from the point of view of the exterior, magnets always appear to have equal and opposite magnetic poles. Because magnetism is fundamentally a product of axial rotation there can be no isolated magnetic monopoles. Magnetism is always dipole.

However magnetic poles are actually an illusion because magnetic field lines continue through the interior of a magnet and just emerge at the other pole in the opposite direction. Thus magnetic field lines always form closed loops as opposed to electric field lines, which radiate outward from electric charges. Magnetic poles are simply a name given to where denser concentrations of field lines enter or exit a magnet.

So the magnetic force is actually along the field lines proportional to their density, which is greater at the entry and exit points. Thus it appears the poles are doing the attracting or repulsing but it's actually the density of field lines where the effect occurs. The opposite poles are due to the field lines pointing inward as they enter at one pole and pointing outward as they exit the other.

Thus magnetism doesn't really have positive and negative poles. It's just a matter of which direction the lines of force are pointing and how dense they are. So for example the magnetic field around a straight current carrying wire has *no poles* because the field lines are all circularly concentric around the wire. So it's the density gradient of the lines that is greater towards the wire that exerts a magnetic force either towards or away from the wire.

Magnetism is different from electricity in this respect, which does always come in positive or negative charges. And positive and negative electric charges are always in different particles.

As with masses, electric charges don't themselves attract or repel but because their electric fields form velocity density gradients in spacetime that produce velocity vectors at every point. As with mass fields the electromagnetic field is an inseparable part of the actual charge and other charges tend to move along velocity vectors up the gradient of the field.

Another source of magnetism is due to the orbital motion of electrons in atoms. This orbital motion produces quite a strong magnetic force but like the spins of most particles it's randomly oriented among atoms and mostly cancels out.

The third form of magnetism is due to the motion of electric charges in currents. When electric charges move through a wire they generate a magnetic field encircling the wire according to the Right Hand grasp rule. When many wires are wrapped tightly in a coil (a solenoid) the magnetic field generated within the coil is multiplied and when the current is properly modulated will rotate an iron rotor. This is of course the principle of the electric motor.

The magnetism generated by particle spin can be easily understood by analogy to that generated by a moving current in a wire. If we slice the spinning particle open along one side from pole to pole and lay it out flat we see that the magnetic field encircles the particle just as it does a wire.

So all three of these magnetic effects are manifestations of electric charges in motion. Magnetism is electric charge in relative motion and this relative motion can be either that of the charges themselves or of an observer relative to them.

An observer at rest with respect to a system of electric charges will see no magnetic field. However if either the charges *or* the observer begins to move the observer perceives it as a current and an associated magnetic field. A magnetic field is simply an electric field seen in a moving coordinate system. It doesn't matter whether the electric field or the observer is moving; all that counts is their relative motion. How this works as an effect of Lorentz contraction along the direction of motion is depicted graphically at (Schroeder, 1999).

So magnetism is actually a relativistic effect of electricity, and electricity is transformed into magnetism by relative motion just as velocity in time is transformed into the mass-energy of velocity in space by relative motion. Both are examples of the Lorentz transform, which is simply the Pythagorean theorem describing the projections of a single vector onto orthogonal coordinate axes. Thus the electric and magnetic fields are 90° orthogonal projections of a single underlying entity just as space and time velocity are.

Thus when we play with a magnet and observe its effects we should realize it works because of the enormous relativistic velocities of electric charges at the particle and atomic levels. The energy within matter is enormous and it's only the nearly exact balance of forces that constrains it in the seemingly ordinary and trivial objects around us. Thus magnetism is a clear everyday example of relativity. Whenever we experience magnetism we are actually experiencing relativity in action.

THE ELECTROMAGNETIC FIELD

Time turns into space with increasing velocity, and electricity turns into magnetism with increasing velocity. And increasing spatial velocity is increasing mass-energy. Thus increasing mass-energy turns time into space and electricity into magnetism.

Now if we just model the electromagnetic lines of forces as consisting of minute helices in accordance with our model of force fields as various forms of intrinsic velocity in space we get a simple attraction and repulsion model of plus and minus poles and charges of these two distinct but interrelated forces.

Electromagnetism is another form of energy and thus according to Universal Reality a form of relative motion, with a strength equal to its velocity density. Like mass electric charges are modeled as spherical fields of velocity densities in spacetime radiating from the center of the charge. These fields of spacetime distortion alter the proportion of space and time distances and velocities at points in the field and the field gradient produces velocity vectors that induce inertial motion in other particles.

Since charged particles also have mass they are associations of two kinds of velocity density, one produced by their mass, and another by their charge. Both these fields are can be visualized as spherical areas of velocity density within a *flat Euclidean* space in which the relative

24

distances and velocities of time and space are shifted at every point in the field as described in the previous chapter.

Both gravitational and electromagnetic fields fall off as the square of the distance due to the simple fact that in 3-dimensional space the area of the surface of a sphere increases by the square of the radius. Thus the strength of fields falls off inversely with distance. Thus the constant total strength of the field over distance is simply diluted by the increasing volume of 3-dimensional space as the distance from the center increases.

Our model of mass and gravitation suggests a similar model for electromagnetism. This is a very neat theory original to Universal Reality with a lot of explanatory power. It also provides an excellent explanation for how the standard theory of electromagnetic fields as virtual photons works.

The difference in the intrinsic spatial velocity of mass and electric charge is in their forms. Electric charges are spherical fields of *minute helical spacetime distortions* in the surrounding dimensional fabric. In other words charges produce a field of miniscule corkscrew twists in the surrounding spacetime that form the field lines of electromagnetic fields and increase the velocity density of points in the field. Of course the actual fields are continuous and fill all space, the field lines are just a graphical sampling of the actual field.

These spacetime distortions produce velocity vectors felt mainly by other charged particles whose own fields strongly couple to them since their helical distortions tend to reinforce or cancel each other out depending on their direction of twist. Electric and magnetic field lines are modeled as separate orthogonal projections of these helical distortions in space.

The transformation of electric force into magnetic force with the spatial velocity of charges occurs as the helical vortices of the electric field begin to tilt orthogonally into the helical vortices of the magnetic field according to the right hand grasp rule. With greater and greater spatial velocity (current flow) the helices tilt more and more and become a magnetic field that appears as field lines of magnetism perpendicular to the electric field lines.

So spatial velocity transforms electricity into magnetism by flipping the individual helical vortices of the electric field towards the perpendicular where they become individual helical vortices of the magnetic field. If the actual velocity of the charges could attain the speed of light all the electric field helices would flip 90° over into magnetic field helices and the electric field would become entirely a magnetic field due to the conservation of the total electromagnetic field.

The helical spacetime distortions are generated by individual charges and rotate in two possible directions, either in the clockwise or counterclockwise direction. These correspond to positive and negative electromagnetic charges or positive and negative magnetic poles. The fact there are only two possible rotational directions for helices neatly explains why there are only two electromagnetic charges and two magnetic poles.

The effective sizes, velocities and density of these helices are all the same since the spins and charges generating them are of equal strength. However the field densities produced by multiple particles are additively scaled to produce the measured values of electric and magnetic forces.

The spins of the charged particles are clearly the source of the helices of their electromagnetic fields, which extend outward from the spinning charges.

While the vibrations of mass come in different amplitudes due to the non-proportionality of particle masses and the additive nature of the gravitational force, the helical vortices of electromagnetism have identical strengths because the strength of their elemental charges are identical. Multiple charges just add to the number of identical helices to produce a denser field. This enables the helices of electromagnetic fields to cancel or reinforce depending on whether they are rotating in the same or opposite directions.

The relativistic effect of linear motion on mass vibrations is a tilting of time velocity into space velocity while the relativistic effect of linear velocity on the electromagnetic helices tilts the individual helices from parallel to the electric field lines towards a perpendicular magnetic field line orientation. The individual helices don't move they just tilt in place. Their projection on the electric field lines at any point is the electric force and their perpendicular projections are the magnetic force and become magnetic field lines.

So the electric force is analogous to velocity in time and the magnetic force is analogous to velocity in space. Both velocity in time and the electric force tilt into their alter egos with linear velocity and the vector sum of each with its alter ego is conserved. This similarity in form is why the electromagnetic force can be modeled as a compacted 5th dimension in Kaluza-Klein theory though our helical model is preferable.

Now helices cancel each other out when they are rotating in opposite directions and reinforce when they are rotating in the same direction. Thus helices rotating in opposite directions cancel where they are pointing in the same direction, and reinforce where they are pointing in opposite directions. And helices rotating in the same direction cancel when they are pointing in opposite directions, and reinforce when they are pointing in the same direction. This is the key to understanding magnetic attraction and repulsion, and the repulsion and attraction of electric charges.

Thus in areas *between* separate poles or charges of the *same* sign the helices cancel each other out and they reinforce in areas *outside* the charges or magnetic poles. Thus the velocity density of spacetime is increased in the areas outside and unaffected in between. Thus the velocity vectors at the points of the poles or charges are directed away from each other and this is the source of the *repulsion* of identical magnetic poles and identical electric charges.

And in areas between separate *opposite* poles or charges the helices reinforce and they cancel beyond producing an area of strong velocity density between the charges so the velocity vectors where the charges are located point towards each other and this is the source of the *attraction* between opposite charges and opposite poles.

In areas external to two opposite charges or poles their helices will be rotating in opposite directions and will almost completely cancel each other out. This is why there is no net electromagnetism in most materials because their collective helices cancel each other out in external areas. Only in materials where slight spatial imbalances of charge polarity exist and are aligned will some residual helices remain. In such cases magnetic effects will be present and the materials will be magnetic.

Thus Universal Reality suggests a simple and elegant new model of electromagnetic attraction and repulsion in terms of the spacetime dilation generated by a helical velocity density just as it did for the similar gravitational effects of mass in the form of simpler vibrations. And the specific forms of the spacetime distortions produced by mass and electric charges neatly explain why there are two opposite electromagnetic charges and only a single gravitational charge.

Both theories are explained in terms of the same spacetime velocity density model, and both are complementary distortions that can exist together in the same spacetime volumes.

Thus both mass and charge can produce their different distortions in spacetime simultaneously. They work together naturally in standard 4-dimensional spacetime. Electromagnetic effects propagate across the vibrational spacetime of general relativity, and gravitational effects are produced by a simpler form of velocity density. Gravitation and electromagnetism are simultaneous distortions or velocity densities in spacetime of different forms. But why the intrinsic velocity of a gravitational field doesn't also tilt the electric force into magnetic force is an open question.

As forms of energy both mass and electromagnetic charges increase the velocity density of spacetime in their specific ways and this affects the relativistic behavior of objects within spacetime but most of the velocity density of electromagnetism consists of helical distortions that strongly couple individually but largely cancel each other out in aggregates of oppositely charged particles.

Thus Universal Reality produces a simple and elegant theory of mass and gravitation, and electric charges and magnetism, as different forms of velocity density that work according to the same underlying MEv Principle that all forms of mass-energy are forms of spatial velocity.

And this model is quite easy to visualize and understand in terms of a flat Euclidean spacetime containing fields of the two forms of velocity density that produce velocity vectors in the direction of slower time. This Neo-Euclidean model is equivalent to the curved spacetime of general relativity, but much easier to comprehend because it directly reflects the flat Cartesian spacetime we actually observe.

Think of each point in this Euclidean spacetime having a fixed c value (speed of light) of combined time and space velocities. The relative motion of either mass or charge just distorts the proportion of time and space velocities so that time slows and distance lengthens. We can still think of the overall spacetime as Euclidean but the relative velocity density of space and time at every point is distorted by mass or electromagnetic charge.

Both the vibrational distortions of mass and the helical distortions of electromagnetism produce spacetime distortions proportional to the relative spatial velocity of their energy. Thus both produce velocity vectors that determine the inertial motion of test particles in their fields that are different depending on whether or not the test particle is charged.

Gravitation is a much weaker force than magnetism, the strength of the relative motion produced is 10^{-36} times weaker than that produced by electric charges, and thus the field densities and vector velocities are much less for a unit of mass than a unit of charge.

However because mass has only one charge (there are no negative masses) masses are all attracted to each other and tend to clump without limit into planets, stars and galaxies and produce very large gravitational velocity density fields.

On the other hand equal electric charges repel each other and cannot clump except in very small units under the influence of the strong force in nuclei. Opposite charges do clump in the form of atoms and molecules, but their opposite helical spacetime distortions almost entirely cancel each other out beyond the clumps when they do, so most of the large scale structure of the universe is due to gravitation rather than electromagnetism.

Thus though electromagnetism is intrinsically far stronger than gravitation, gravitation rules on cosmic scales and electromagnetism mainly just holds atomic matter together with external effects largely cancelling out. There are large magnetic fields on cosmological scales but their strength is generally much less than gravitational fields.

As a form of energy the helical spacetime distortions of electric charges do have some gravitational effect on uncharged particles as predicted by the stress-energy tensor of the Einstein field equations, but these are generally negligible compared to those of mass. Since charged particles are normally paired the helices largely cancel in areas external to the particles and the velocity density between closely adjacent points will be minimal. Thus the velocity vector will be relatively small compared to the force on a coupled charged particle, which is effectively canceled in one direction and doubled in the other.

This is how our velocity density model addresses the initially vexing question of why the gravitational force is so much weaker than the electromagnetic force but has such a larger relativistic effect.

The gravitational force is due to the vanishingly small difference in velocity density on either side of a test particle along an axis towards its source mass. Thus the resulting velocity vector towards the gravitating mass is also extremely small. However the helical velocity densities of the electromagnetic force couple to those of other charges and so completely cancel on one side and completely reinforce on the other along their common axis. This effectively doubles the *entire value* of the velocity density at the location of the particles. The difference in gravitation's small *gradient* in velocity density at the particle scale and electromagnetism's doubling of the *entire* velocity density is enormous. Thus the electromagnetic force is intrinsically much stronger than the gravitational force.

The velocity density produced by electromagnetism is considerable, but since charges are normally paired ambient velocity densities effectively cancel and the difference on proximate and opposite sides of *uncharged* test particles is effectively nonexistent so the gravitational effects of electromagnetism will be vanishingly small. Note also that mass produces a much greater gravitational effect than energy since by $E=mc^2$ the equivalent amount of mass m in a unit of energy is E/c^2 which is an extremely small number.

Another difference between electromagnetic and gravitational fields is that electromagnetic fields can be shielded but gravitational fields can't be. Again this is due to the fact that the helical velocity densities of electromagnetism come in two opposing rotations. Thus it's generally possible to construct a shield that either damps or diverts an electromagnetic field but adding any shield made of mass or energy just adds to a gravitational field rather than blocking it.

Thus our model of electromagnetism seems consistent with standard scientific theory and explains its basic concepts quite well in terms of velocity densities just as it explains mass and gravitation and the conservation of mass-energy. Thus Universal Reality demonstrates that both gravitation and electromagnetism are manifestations of the single underlying principle that all forms of mass-energy are forms of spatial velocity.

PHOTONS

Photons of electromagnetic radiation are now easily understood as units or quanta of the electromagnetic field that break away from the charges that generate the field and fly off on their own taking some of an electron's orbital energy with them. Their energy can also be absorbed by electrons and kick them into higher energy orbitals. Photons are essentially free quanta of electromagnetic fields that either break away or are absorbed back into fields in the process of emission or absorption of orbital energy. Likewise *gravitons*, if they exist, would be individual vibrations of gravitational fields that have broken away from the field to become actual rather than virtual.

Thus ejected photons are units of orbital velocity converted into the helical waveform velocities of the electromagnetic field. This is an excellent example of how the conservation of mass-energy always involves the conversion of one form of spatial velocity into another.

The emission of photons of electromagnetic radiation is the result of the acceleration of an electric charge(s). This can occur either due to a free electrical charge or system of charges changing direction as in an antenna, or when an electron decelerates to a lower orbital and emits its lost orbital energy as a photon of electromagnetic energy.

Because they are quanta of helical electromagnetic fields photons can be described as helical vortices in space just as electromagnetic fields are, but in the form of localized packets of the field moving through space at the speed of light. The helical packets of photons are no longer bound to a particle charge and thus fly off at the speed of light. By the STc Principle photons always travel at the speed of light because they have no internal structure to be computed and thus no internal proper clock time velocity. Thus their total c valued spacetime velocity is all spatial velocity.

When a mass or charge is present its velocity density effects propagate across spacetime at the speed of light. This is because all computational effects propagate at the speed of light. Remove the mass or the charge and the surrounding velocity density field dies off at the speed of light.

Thus if an individual helix breaks away from a charge it naturally propagates through spacetime at the speed of light. Photons of electromagnetic radiation are just individual electromagnetic helices freely propagating through space at the speed of light because they aren't anchored to a source charge.

Photons are individual helical packets moving linearly through spacetime as opposed to spherical fields of helical velocity densities surrounding charges. They carry energy proportional to the frequency of their helical rotations. Thus the helical spacetime distortions around

31

charges can be considered as clouds of virtual photons and they produce energetic effects when interacted with. Photons take the form of alternating orthogonal electric and magnetic waves. Together these oppositely oscillating waves take the form of a rotating helical spacetime distortion traveling at the speed of light.

Thus Universal Reality naturally explains light as a form of electromagnetic energy and naturally explains why it moves at the speed of light through space. The speed of light is actually the intrinsic velocity of spacetime, the maximum speed of clock time, and the speed at which computational effects propagate through spacetime, and light just happens to propagate at the speed of light because it has no proper clock time rate.

The helical distortions in spacetime surrounding charges have fixed amplitudes, frequencies and wavelengths since the basic units of electrical charge and spin that produce them have fixed strengths. They differ only in the direction of their helical twists corresponding to the sign of the charge that produces them.

However electromagnetic radiation isn't anchored to a fixed charge so its helices can be produced in a more or less continuous spectrum of frequencies proportional to the amount of relative motion converted to produce them. Thus the energies of photons range from radio waves through visible light to gamma rays.

Like the helices extending from charges, those of electromagnetic radiation can also rotate in either a clockwise or counterclockwise direction. This accounts for the clockwise and counter clockwise circular polarization of light. In most cases, such as the light from the sun, light beams are a mixture of the two polarizations.

The helical waves of photons generally don't cancel each other out or cause attraction or repulsion because they tend to be mixtures of many different frequencies and due to their great velocities effects tend to be more or less instantaneous and immediately cease. However coherent beams of photons of the same frequency such as those emitted by lasers can interfere if correctly tuned (Wikipedia, *Laser)*.

We are now at a point where we can integrate elementary particles and the quantum world into Universal Reality in a manner that accomplishes the long sought unification of relativity and quantum theory.

QUANTUM REALITY

UNIFYING RELATIVITY & QUANTUM THEORY

The next piece we need to add to the puzzle is how to incorporate quantum reality in a form that's consistent with general relativity into our theory. The key here is rethinking the fundamental source of their apparent incompatibility.

The fundamental reason relativity and quantum theory appear incompatible is they are built on incompatible models of spacetime. Both assume that spacetime is a pre-existing physical container for events but the spacetime of quantum theory is passive and fixed while that of relativity is dynamic and affected by its contents.

The solution is to understand that what we call spacetime isn't a pre-existing physical container at all but instead is *created by quantum events* in the form of the universal network of dimensional relationships produced in satisfying the conservation of energy as particles interact.

This enables quantum events to construct a spacetime that is compatible with relativity by scaling the dimensionality of the spacetime created by the amount of mass and energy of the interacting particles. This scaling is imperceptible at the quantum scale so the equations of quantum theory continue to work as before, but they show up at the relativistic scale of large aggregates of particles. In this manner particle interactions create a spacetime compatible with both relativity and quantum theory without negatively affecting either.

RETHINKING SPACETIME

While the concept of spacetime being created by quantum events may initially seem counter intuitive it's actually quite logical and obvious when we objectively consider the fundamental nature of spacetime itself.

We think of spacetime as being a physical container in which individual objects occupy locations separated by empty space. But the problem is we never actually observe any such empty space and the existence of anything that can't be observed is automatically suspect.

All we actually observe in our sensory and scientific observations and measurements are the *dimensional relationships* among objects especially those relative to us. We never ever observe empty space itself, as emptiness is quite obviously not anything observable. We think we observe objects *in* empty space but again all that we actually observe are dimensional relationships *among* things, cued by apparent versus actual sizes, directions, and orientations relative to ourselves and other objects.

As we will see later in the chapter on *The Simulation* it's actually our own brains that first generalize the observed dimensional relationships among multiple things into a single consistent mental mapping, and then reify this mental mapping into the concept of a 'physical' spacetime in which it can mentally place subsequent objects and events. So what we think of as a physical spacetime is actually an information construct in our mind's simulation of reality.

So what actually exists in the external world is not an encompassing physical spacetime but the individual dimensional relationships among particles and particulate objects produced by the conservation of energy and momentum as they interact.

Thus what we think of as a preexisting all encompassing empty physical spacetime exists only in our brains as a conceptual model rather than in the external world. It's an adaptive mechanism that has evolved to help us to make better sense of our surroundings by mentally placing all observed dimensional relationships in a single consistent dimensional mapping in our minds. This greatly simplifies our internal simulation of reality and enhances our survival by making the world around us much easier to understand.

So it turns out there aren't any logical or scientific reasons to believe in spacetime as a pre-existing physical container when the dimensional relationship model is reasonable and consistent and has the huge advantage of unifying relativity and quantum theory as well.

Not only that but as we will see this model also immediately resolves all the seemingly paradoxical aspects of quantum reality because all the apparent paradoxes of quantum theory are *only with respect to a pre-existing spacetime container which no longer exists*!

So by understanding how quantum events create spacetime we can both unify quantum theory and relativity and simultaneously resolve all apparent quantum paradoxes.

And by understanding that spacetime is a computational structure created by particle interactions we also open the way to better understanding the entire universe as a computational system, which quickly leads to even more revelations and simplifications.

HOW PARTICLE EVENTS CREATE SPACETIME

How particle interactions create spacetime in a manner that unifies relativity and quantum theory is fairly easy to understand when the process is carefully explained. An overview that should make the basic process clear is presented in this chapter. Additional details are explained in *Unifying Relativity & Quantum Theory* (Owen, 2016).

The universe consists entirely of elementary particles and their interactions. All the familiar classical scale objects and processes are *emergent manifestations* of particle interactions in aggregate. Particle interactions include both those of free particles and particles bound in atoms and molecules and the materials they compose.

Particles can be thought of as composed of small sets of *particle components* that are conserved, with very few exceptions, through all particle interactions. These particle components include baryon (quark) and lepton number, mass and the other force charges, spin, and space and time parity. These particle components can be considered the ingredients of particles that make them what they are.

Because particle components are conserved through all interactions even when particles change into one another indicates that the particle components, rather than the elementary particles, should be considered the true fundamental components of the universe. Particles aren't conserved but particle components are.

Each particle component is conserved separately. This means that the sum total of each particle component exiting any interaction must equal the sum total of that same type of particle component that entered the interaction. Thus particle interactions can be considered as redistributions of the same particle components that entered the interaction.

Most particle component values come in a few small values that can be either positive, negative or zero. It's the sum of the *values* of each particle component that are conserved independently rather than the total *number* of particle components of each type. For example a high-energy collision of two massive particles can produce a shower of new particles containing many more particle components than the colliding particles contained. However the additive sum of each particle component type before and after must remain the same.

Total mass-energy is also conserved through every particle interaction. Since all types of mass-energy are different forms of spatial velocity this means that total spatial velocity is conserved through all particle interactions. Thus particle interactions often result in the conversion of some of the total mass-energy into other forms of mass-energy. For example some of the original mass may be converted into the kinetic energy of linear velocity as happens in nuclear explosions.

A set of multiple new particles is typically produced by a particle interaction and the particle components and total mass-energy are redistributed among these particles. What this means is that all the individual particles produced by any particle interaction are necessarily interrelated on their individual particle components and mass-energy values because the totals across all emitted particles must equal the totals entering the interaction.

For example if the conserved total mass-energy of an interaction is E, the total mass-energy of all the particles emitted by the interaction must sum to E. The mass-energy of each individual particle will be the sum of its mass, linear velocity kinetic energy, wave frequency energy etc. but the sum total of the mass-energies of all emitted particles must equal E.

What this means is that all the mass-energy values of the emitted particles are all interrelated. Since they must all sum to E, the individual mass-energy value of each individual particle will be related to the mass-energy values of all the others.

This is also true of each type of particle component individually. Each type is individually conserved so that each emitted particle is interrelated to all the other emitted particles on each of its particle component types. We say that all the particles emitted by any particle interaction are *entangled* on their mass-energies and each of their particle component types. *Entanglement* is the interrelationships among the individual particle component types and mass-energies of the particles resulting from any particle interaction due to the conservation of mass-energy and particle component types.

Contrary to popular belief entanglement is an easy to understand fundamental corollary of conservation that applies at all scales. For example if we break a cookie into two pieces, the two pieces will be entangled on their weights and volumes because the sums of the weights and volumes of the two halves must equal those of the original unbroken cookie.

And if the cookie contained chocolate chips and pecans the two halves will be entangled on chocolate chips and pecans independently because the total in the two halves must equal the number in the original cookie. If we count how many are in one half we know how many are in the other half without counting. The same holds for every ingredient of the cookie. The two halves resulting from the break will be entangled on every one of their ingredients.

Entanglement works almost exactly the same at the quantum scale. All the individual ingredients of the resulting particles must equal their total amounts in the original particles entering the event. The only real difference is how entanglement is measured at the quantum scale and the implications that has for how spacetime is created from those measurements.

Each particle interaction is computed by a separate application of the elemental program and the universal processor. Thus all the particles emitted by an event are computed by a single elemental program application that continues to compute all the interrelated particle components together to preserve their conservation. Because all the emitted particles are being computed together as a single process this is referred to as a *coherent process*.

This same application continues to compute all the emitted particles as a single group until any undergo subsequent interactions at which point a new application of the elemental program and processor begins to compute that new interaction and its emitted particles as a new coherent process.

As time progresses particles continue to interact and chains of interactions build up and join in vast networks of interactions. In fact it's reasonable to assume that all the particles in the universe are connected in a single network of interactions dating back to the big bang. Thus all the elementary particles in the universe become interconnected in one vast *entanglement network* that interrelates them all on their individual particle components and mass-energies.

Because the mass-energies of all particles in the universe are interrelated they all have dimensional relationships to each other and the network of all these dimensional relationships is logico-mathematically consistent. There is a single internally consistent logico-mathematical network of dimensional relationships among all the particles in the universe that is produced by the conservation of mass-energy in their interactions.

Observers are particulate structures that are part of this entanglement network. Thus all our observations are part of the overall dimensional consistency because all observations reduce to particle interactions. It's this dimensional inter-consistency of observations, both scientific and sensory, that humans misinterpret as an all-pervading empty spacetime within which events occur.

The human mind simplifies its conceptual model of the actual information universe in a number of ways as explained in the chapter on *The Simulation*. In this instance it takes the inter-consistency of its observations of the dimensional interrelationships among objects and mentally creates a continuous all pervading imaginary spacetime from those dimensional points. It then reifies this mental construct into the illusion of a physical spacetime centered on itself. This is one of the fundamental illusions of human existence, the illusion that we live in a continuous physical spacetime. But this is actually a mental extrapolation of sensory observations of dimensional relationships among individual perceptual objects. By placing these dimensional points within a 3-dimensional mental map we assume there is an empty physical space between those points that doesn't actually exist.

It's important to understand that the entanglement network is not something that actually exists as an independent data structure in the universe. It's entirely a *conceptual structure* representing how informed observers understand the nexus of particle interactions. All the actual data of the universe are those of the particles and particle component values themselves. Only when considered in aggregate by an observer are they seen to be interrelated in an entanglement network.

In this conceptual overview the entanglement network represents the entire data structure of the history of the universe. However only the current present moment slice of the entanglement network represents the actual universe as it currently exists in the universal present moment it creates by its presence.

This model neatly explains how particle events actually create spacetime rather than occurring in an already existing empty spacetime. It's a subtle melding of how particles interact and how our minds work.

But we still need to explain how this model unifies relativity and quantum theory and how it explains quantum indeterminacy and resolves all quantum paradox. This can be explained by understanding how separate applications of the universal processor actually compute particle interactions.

QUANTUM INDETERMINACY

The quantum world is characterized by dimensional indeterminacy as seen in wavefunctions, the uncertainty principle, and the zero-point energy. This can be easily explained by a simple tweak to how the processor that computes the universe operates.

We've already seen that the STc Principle that underlies relativity can be explained by the processor allocating a fixed number of cycles to the calculation of the total spacetime velocity of everything it computes. The total number of cycles always results in a fixed c speed of light velocity of every process in the universe. If cycles are allocated to computing any velocity in space, the number of cycles allocated to computing velocity in time is reduced. This is the computational source of all relativistic effects.

Now if the allocation of space versus time cycles randomly oscillates at the quantum scale all the dimensional indeterminacy of quantum processes can be explained. There will be a separate random oscillation pattern for each application of the elemental program as it computes the evolution of each coherent process where a coherent process consists of all the particles emitted by any event prior to any subsequent interactions.

All the particle components and mass-energy of an event are exactly conserved as before. However when the mass-energies of emitted particles are manifested dimensionally as positions and velocities their space and time values will be slightly conflated by the random oscillations. This results in a slight intrinsic uncertainty in the positions and velocities of the particles.

This is because all the dimensional indeterminacy of quantum phenomena can be reduced to a slight random conflation of space and time velocities. It's as if the metric against which velocity and position are being measured isn't quite exact but continually oscillates between space and time making the measurements appear fuzzy.

But since physics assumes the spacetime metric is exact and fixed, though it's actually oscillating, all measurements with reference to it will appear to randomly oscillate. As a result space and time values will be conflated at the quantum scale. In particular there will be a Heisenberg uncertainty in the measurement of particle positions (space) and momentums (time).

These spacetime oscillations are also the reason why particle trajectories appear to take the form of wavefunctions because wavefunctions too are characterized by random oscillations in the positions and velocities particles probably have. Thus wavefunctions are essentially the same as our processor oscillation patterns.

Since the set of particles emitted by any event is computed as a separate coherent process by its application every set of coherent particles will have its own unique random oscillation pattern and thus be dimensionally indeterminate with respect to all other particles. Every set of coherent particles will have its own unique random oscillation pattern and thus every coherent particle set will be dimensionally indeterminate with respect to all others at the quantum scale. This is the computational source of all quantum indeterminacy.

However since all the particles in a coherent set are being calculated by a single application with a single oscillation pattern their dimensionalities continue to be exact with respect to each other until one is measured and as a result becomes part of a new coherent process with a new oscillation pattern. (The wavefunctions of coherent particles can differ because there are other components that determine the form of wavefunctions, but the underlying oscillation patterns should be identical.)

Thus when an observer measures the dimensionality of a particle it will be indeterminate with respect to his own at the quantum scale because the particle and observer are being computed by separate processor applications. This explains the fundamental source of quantum randomness.

In essence the continuously oscillating random dimensionality of each coherent set of entangled particles can be considered as a separate *dimensional fragment*, since it embodies the independent dimensionality of the particles in that set which is indeterminate with respect to the dimensionality of other coherent processes including those of an observer or measurement apparatus.

Thus what we think of as spacetime is actually constructed from myriads of separate mutually indeterminate dimensional fragments produced by the conservation of mass-energy in particle interactions. But this occurs only as separate dimensional fragments are linked together by subsequent decoherence events.

DECOHERENCE

Now a measurement is ultimately a particle interaction and for particles to interact all their particle components and mass-energies must have exact values so they can be conserved. If they didn't have exact values they couldn't be conserved through the interaction and exactly redistributed among the emitted particles.

But the problem is that particles from different coherent processes are dimensionally indeterminate with respect to each other. So for particles to interact their dimensional indeterminacy with respect to each other must be resolved to exact values so that mass-energy can be exactly conserved. This resolution of dimensional indeterminacy to exact values is called *decoherence*.

Thus when particles interact in measurements or otherwise the interaction selects exact position and velocity values from the random possibilities expressed by their oscillation patterns (wavefunctions). Thus exact values are selected but randomly because the oscillation patterns of the interacting particles were indeterminate with respect to each other.

This process of mutually selecting exact dimensional values from the interaction of oscillation patterns or wavefunctions is called *decoherence*. Whenever particles interact their dimensional indeterminacy with respect to each other is resolved in exact values randomly selected from their wavefunctions. Only thus can an interaction that conserves mass-energy occur.

Decoherence is always a mutual process because the interaction of particles simultaneously produces exact dimensional values for both as it must for their total mass-energies to be exact and able to be conserved. Essentially the interaction of the two wavefunctions or oscillation patterns randomly selects exact dimensional values from the possibilities carried by both.

When two particles from separate coherent processes decohere their coherence with their previous dimensional fragments is lost. They drop out of their previous dimensional fragments and become part of a new coherent process in a new dimensional fragment. This new dimensional fragment is computed by a new application with a new oscillation pattern that takes the form of new particle wavefunctions that are indeterminate with respect to those of all other dimensional fragments.

As a result the entanglement network can be envisioned as a universal network of event nodes, where interacting particles have exact particle component and mass-energy values, connected by lines representing particles whose dimensionalities are indeterminate with respect to each other at the quantum scale. The entanglement network is a network connecting all particles through their interaction events. Particles have mutually exact dimensional values at the event nodes where they interact but the nodes are interconnected by wavefunctions dimensionally fuzzy with respect to each other.

It's the dimensional consistency of this alternately exact and fuzzy entanglement network that we human observers interpret as a dimensionally fuzzy spacetime at the quantum scale and an exact relativistic spacetime at the classical scale.

THE ENTANGLEMENT NETWORK

We know that wavefunctions and the Schrödinger equation describe the quantum world with great accuracy. So our processor oscillation model must exactly duplicate the operation of wavefunctions to be correct. And it has to do this as it simultaneously explains how spacetime emerges from particle interactions rather than being a preexisting physical structure.

But even in a classical world devoid of quantum effects it's easy to understand how particle events create spacetime. So this will be explained first and then quantum randomness in the form of processor oscillations will be added to complete the picture.

1. In the initial classical version the entanglement network is a completely exact structure and all particle dimensional values are exact. All particles have exact dimensional relationships to each other.
2. The entanglement network will consist of exact conservation events linked by particles moving along exact dimensional trajectories.
3. The conservation of energy-momentum in events results in particles moving along exact trajectories with exact positions and velocities in every P-time tick.
4. However all these values are numeric data in a 3-dimensional matrix in computational space rather than physical particles in an actual physical spacetime. It's exactly the same as a virtual reality program. All the dimensional data actually exists in a data matrix in the controller's memory, and is then transferred to the headset where it's projected into the appearance of a 3-dimensional world.
5. In the actual world all the data of the observable universe similarly exists in a 3-dimensional data matrix in computational space, and our mind simulates it and projects it out around us as the appearance of our familiar 3-dimensional world.
6. So the observable universe is just a much more convincing virtual reality program where computational space is the controller and our simulation is a micro controller that projects the data of computational space into the semblance of a physical 3-dimensional universe.
7. So the entire observable universe exists only as particle data in computational space and we are part of that data.
8. In aggregate the data of all particles and their interactions in computational space takes the emergent form of the entanglement network.

9. The entanglement network consists of the nodes of all particle events connected by lines representing all the particles in the universe moving through time from event to event.

10. The entanglement network is an emergent structure recognizable only by observers with simulations able to store and compare the actual particle data so as to be able to recognize relationships among particles and particulate objects.

11. Conceptually the entanglement network contains the entire computational history of the observable universe but in actuality it's the emergent structure of only the current data state of the observable universe in the present moment as it's continually recomputed.

12. What we call dimensional spacetime is the logico-mathematical consistency of the dimensional aspects of the entanglement network. It's constructed particle interaction by particle interaction and consists entirely of all current individual dimensional relationships among particles.

13. The entanglement network is an emergent structure that contains all the data of the observable universe. The logico-mathematical consistency among its individual dimensional relationships is what our simulation presents to us as a physical spacetime.

14. But our simulation of a physical spacetime populated with physical objects exists only as neural data in our brains. It's a dynamic data model sampled from the enormously more complex entanglement network that itself exists entirely as data in computational space.

15. Thus our simulation is a little subroutine of the universal program that computes the entire observable universe that continually recomputes our view of the observable universe.

16. Both we and our simulation and the universe itself are all running programs computing their respective data structures. We and our simulation are subroutines running with the universal program that continually recomputes the entire observable universe.

17. So it's clear how even a classical non-quantum spacetime is actually a data structure created by aggregate particle interactions rather than being a preexisting physical container.

Next quantum reality can easily be added to this model by incorporating the processor oscillations in a manner that exactly reproduces how wavefunctions function as spacetime emerges. That creates a spacetime that displays quantum randomness at the particle scale and relativity at the classical scale.

1. To extend our model to quantum phenomena all that's needed is to have the processor oscillations compute each particle's positions and velocities around its exact trajectory. This duplicates how wavefunctions work because they too express the probabilities of the random positions and velocities a particle may be found in around the trajectory of its wavefunction. Wavefunctions also have exact trajectories and time evolutions. It's only where particles may be found within their wavefunctions that's probabilistic.
2. Wavefunctions are mathematical structures that have a real and an imaginary part. The real part has the form of a sine wave. The imaginary part is the probability density of where a particle may be found within the wave.
3. So to have the oscillations mimic wavefunctions they must have two parts. An overall oscillation pattern and a random selection of dimensional value(s) within that pattern at every P-time tick. The random selection is the specific conflation between space and time velocities at each tick of the coherent process computing it.
4. Though this is computed in computational space it can be visualized as an oscillation pattern that looks like a wavefunction moving along an exact particle trajectory. At each point in time a dimensional value is randomly selected from the oscillation pattern. Just as with wavefunctions the selection is random but the value selected is exact.
5. In our oscillation model a random value is selected in every P-time tick. This is effectively a *virtual decoherence* that allows the particle's interaction with any fields to be computed and to test for an *actual decoherence* with other particles.
6. If only a virtual decoherence occurs the exact trajectory may be affected by its interaction with the field. In this case it may remain part of its original coherent process depending on the strength of its interaction with the field and whether it produces an actual decoherence or not. There can also be partial decoherences not covered here. See *Universal Reality* for more (Owen, 2016).
7. Even if particle trajectories meet in time an interaction may not occur if the random virtual decoherence values of the particles don't sync. This explains phenomena like quantum tunneling where particles occasionally appear to pass through solid objects.
8. If two particles virtually decohere close enough to each other then a particle interaction occurs and the virtual decoherence becomes an actual decoherence. (Particle interactions are actually more complex than being in the same position at the same time but this will suffice for now.)

9. If an actual decoherence occurs both particles abandon their previous exact trajectories and relocate to their mutual decoherence location and the computation of a new particle interaction event is initiated.
10. The interacting particles also drop out of their previous coherent processes and become part of a new coherent process computed by a new application with a unique new oscillation pattern.
11. As they drop out their entanglements with the other particles in their previous dimensional fragments are lost and all particles exiting the new event become entangled as a new dimensional fragment is formed.
12. Because all particles emitted from an event are computed together by the same application with a single oscillation pattern the entanglement of all the emitted particles is maintained as they evolve along their trajectories. Their common random dimensionalities are all related and vary together to maintain conservation.
13. This explains the spin entanglement paradox because the equal and opposite spins of the two emitted particles are continually being computed as a single coherent process so even when they are randomly selected at each tick they are always selected as an equal and opposite pair. This continually maintains the conservation of spin. This differs from traditional quantum theory, which has no explanation for the spin entanglement paradox.
14. All the dimensional values of all particles emitted by an event are coherently entangled in each P-time tick. (This is a potential test of the theory though it may be impossible to confirm it due to the likely impossibility of simultaneous measurements at the resolution of P-time ticks. Oscillation patterns rather than wavefunctions must be tested because the form of wavefunctions depends on the type of particle as well as their common oscillation pattern.)

With the addition of oscillation patterns to incorporate quantum phenomena the resulting entanglement network now looks like this.

1. The entire observable universe exists as data in the computational space of the quantum vacuum.
2. The only actual data in computational space is the data of the elementary particles and their current particle component values, which is the data of the entire observable universe.
3. This is stored in the form of 3-dimensional data arrays that are continually recomputed at each P-time tick.

4. In aggregate all this particle data takes the form of an entanglement network. The entanglement network is an emergent data structure similar to how a meaningful moving picture emerges from aggregates of continually changing individual pixels on a TV screen. It's meaningful only to properly configured observers.
5. The entanglement network consists of event nodes connected by particle trajectories.
6. Events exactly conserve all particle component values including energy-momentum values independently. The totals exiting the event equal the totals that entered the event.
7. The particle trajectories connecting event nodes are exact as well. However the actual positions particles are found along their trajectories is probabilistic because they are computed by applications that conflate space and time velocities at the quantum scale.
8. However all the particles exiting a single event are entangled on each of their particle component and energy-momentum values because they are all being computed by the same application with its own unique oscillation pattern. This maintains the conservation and entanglement absent one of the particles interacting with another particle.
9. Thus the dimensionality of particles in separate coherent processes is indeterminate with respect to each other and each coherent process is effectively a separate dimensional fragment.
10. If particles from separate dimensional fragments randomly decohere in sufficient proximity they interact in a new particle event. At this point they leave their leave their previous trajectories and relocate to the point of their mutual decoherence. They also leave their previous coherent processes and become part of a new coherent process that computes the event and all the particles it produces with a new oscillation pattern.
11. Thus coming into events there are little quantum scale discontinuities in the exact dimensionality of the entanglement network where particles assume new exact dimensional values as they mutually decohere. This is the source of the dimensional indeterminacy that characterizes quantum scale phenomena.
12. Nevertheless the dimensional consistency of the entire entanglement network is maintained at the classical scale and relativity continues to work as before.

Thus our oscillation pattern model incorporates both quantum theory and relativity in a single unified model by explaining how spacetime is computed by particle events rather than being a pre-existing physical structure within which events occur.

HOW PARTICLE EVENTS COMPUTE RELATIVITY

Thus as particles interact they create the dimensional relationships among them our minds interpret as a dimensional spacetime. Individually particle interactions obey the STc Principle because the universal processor that is its source computes them. Thus all the dimensional relationships produced by particle interactions are scaled by the STc Principle and are compatible with general relativity.

However these relativistic effects are largely imperceptible at the quantum level so the laws of quantum interaction appear unchanged and work exactly as before. But at large dimensional scales these computations at the particle level produce all the effects of general relativity.

Thus we have a straightforward explanation of both how particle interactions create what we call relativistic spacetime and how the relativistic spacetime they create determines how they move. The dimensional consistency of the entanglement network we interpret as spacetime is scaled by the STc Principle as it's computed, and the intrinsic velocity density fields produced affect the trajectories of particles around which their dimensionalities randomly decohere. This is equivalent to the adage that 'mass tells spacetime how to curve, and spacetime tells mass how to move'.

Both general relativity and quantum theory are produced by the way the universal processor computes particle interactions. It computes all relativistic effects by allocating a fixed c value number of processor cycles to first computing spatial velocity and those left over to computing the internal evolution of processes which manifests as their clock time rates. But there is also a minute random oscillation in the allocation of space versus time cycles in all computations. It's these random oscillations that produce all the quantum indeterminacy at the particle scale.

Thus the same computational process simultaneously computes both general relativity and quantum theory. These computations occur not in a preexisting spacetime container but in the computational space of the quantum vacuum. Each individual computation produces dimensional relationships among the particles involved. The aggregate result of all such computations is the dimensional consistency among all particles and particulate objects that takes the emergent form of the entanglement network. What we call spacetime is the dimensional consistency of the entanglement network. Our minds simulate this dimensional consistency as an encompassing physical spacetime, project it out around us, and populate it with objects based on their observed dimensional interrelationships.

Thus by replacing the notion of spacetime as a preexisting physical structure that's treated differently by relativity and quantum theory we unify the two theories by recognizing how spacetime is actually computed by particle events. Now quantum events produce dimensional relationships among particles that obey quantum equations at the particle scale and the equations of general relativity at the aggregate scale.

This is an enormously important paradigm shift that effectively unifies the two most important theories of modern physics. This in turn opens the way for important additional insights into the nature of reality and how it works.

Note that in standard general relativity all frames are treated as equally valid, no frame is absolute, and there can be inconsistencies among frames. This might appear inconsistent with a single unique universal entanglement network. However the solution lies in the insight that there must be a single universal reference metric to explain the absolute nature of rotation and motion along world lines. This is the computational metric in which the observable universe is computed which corresponds to the entanglement network. All observer frames and any incompatibilities they may have are computed as views within this universal reference metric and so this single metric maintains universal consistency. This is explained in detail in the section on The Universal Spacetime Metric.

RESOLVING QUANTUM PARADOX

The beauty of the insight that particle events create spacetime is that it also immediately resolves the apparently paradoxical nature of quantum processes because quantum processes are only paradoxical with respect to a fixed preexisting dimensional spacetime that no longer exists! This is strong additional evidence Universal Reality is likely correct.

Consider for example the classic example of the entanglement of the spins of two particles produced by a single event. In the usual example they are produced so they have equal values but opposite orientations.

1. Because the spins are particle components produced by a single event their values are exactly conserved by the event.
2. This entangles the spins so that their orientations are created exactly opposite to each other.
3. Because the evolution of the two particles from the event is being computed as a single coherent process their spin orientations will continue to be exactly opposite as they travel away from the event on whatever trajectories they take.
4. However this exact mutual orientation will be indeterminate with respect to other processes including a measurement apparatus because it was created randomly without reference to any of those processes. Their mutual orientation is being computed by a separate processor application with its own random oscillation pattern.
5. Thus the mutual spin orientation of the two particles could be in any possible orientation with respect to the measuring apparatus though they always remain opposite to each other.
6. However when the spin of one particle decoheres in a measurement then the exact mutual orientation of the two spins becomes aligned with that of the apparatus and a subsequent measurement will always find the orientation of the other spin to be opposite to that of the first.
7. It's important to understand there is no actual change in the opposite spin orientations of the two particles when they are measured. It's simply a matter of aligning the dimensional fragment of their opposite spins with that of the laboratory that measures them.

Prior to a measurement the dimensional fragment of the two opposite spins existed as a completely independent spacetime fragment that had no orientation relationship to any other or to any fixed reference space since that doesn't exist. This is because the only thing actually computed in the event was the orientation of the spins with respect to each other not to anything else.

Thus the first measurement simply aligns the orientation of the two spin dimensional fragment with that of the laboratory through a common decoherence event. And that's all there is to it. There is no 'faster than light' transmission of spin orientations between the particles because their already fixed mutual orientation doesn't actually change. It's just randomly aligned with the orientation of another spacetime by becoming part of that spacetime by linking to it in a decoherence event that joins them.

When completely separate spacetimes or spacetime fragments are independently created they are completely independent in all respects and have no correspondence of scale, orientation, positions, or anything else. Unless they both exist in a common background reference spacetime which dimensional fragments don't, they are completely isolated and unknown to each other. However if particles from these separate spacetime fragments interact their dimensionalities automatically align and become part of a single spacetime without either spacetime actually changing.

This is the solution to the apparent spin orientation paradox. It appears paradoxical in the context of a preexisting spacetime that encompasses both the particles and the observer but when spacetime is recognized as consisting of initially independent dimensional fragments it's not paradoxical at all.

Physicists are continually troubled by the apparent 'non-locality' and 'faster than light' communication between the two spin particles but these concerns both vanish when properly understood.

Since the whole process is all being computed in computational space rather than in a physical spacetime there is no reference spacetime for the spin measurement process to be 'non-local' with respect to. Every quantum computation is inherently non-local because it's creating spacetime rather than taking place within it.

And there is no faster than light communication from one particle to another because their mutual spin orientation never changes. The measurement just randomly aligns the pre-existing orientation with the orientation of the laboratory in which it's measured.

All other apparent quantum paradoxes are similarly resolved and become natural and non-paradoxical when understood in this context. Examples of how other types of paradox are resolved are given in *Unifying Relativity & Quantum Theory* (Owen, 2016).

Thus having quantum interactions create spacetime rather than taking place within it automatically resolves all apparent quantum paradoxes because quantum processes seem paradoxical only with respect to a fixed preexisting spacetime that doesn't actually exist!

THE ILLUSION OF SPACETIME

Thus what we call spacetime is an illusion. It isn't the all-encompassing preexisting dimensional physical structure we imagine it to be. It's actually the internal consistency of the dimensional relationships among interacting particles and particulate objects including those of our perceptual system.

Thus particle interactions can be said to create spacetime but it's important to understand that spacetime is not an actual thing with an actual existence. It just our mental interpretation of the dimensional relationships among the particle interactions we participate in.

All that actually exists is the data of particles and particle components in the non-dimensional computational space of the quantum vacuum. Particles and their components are the only data that exists within this computational space. And what we think of as spacetime is simply an imaginary dimensional edifice our minds construct from the numeric relationships among particles in this computational space to make better sense of it.

So what we call dimensional spacetime is at best an emergent structure that emerges from particle interactions in aggregate, and as an emergent structure is visible only to observers properly tuned to detect it. One of the greatest accomplishments of living organisms is that they have become aware of the emergent structures of the entanglement network including spacetime and developed simulations that model it as an encompassing physical entity that enable them to plan and execute dimensional actions with respect to other particulate objects.

However the internal consistency of this emergent spacetime is far from exact at the quantum scale. At the quantum scale individual particle events create dimensional fragments that are inherently indeterminate with respect to each other.

Though the dimensionality of these dimensional fragments is always at least partially constrained by that of the events that create them some of their aspects can be completely independent spacetime fragments. For example mutual spin orientations can be a completely independent spacetime fragment as explained above.

Thus the logico-mathematical edifice of spacetime consists entirely of fuzzy dimensional fragments that transiently decohere to produce exact dimensional points only to subsequently evolve as new dimensional fragments fuzzy with respect to each other and with respect to human observers.

But also because it's continually being created by particle interactions spacetime depends entirely on particle interactions for its continuing existence. If particle interactions cease, spacetime instantly vanishes except as its data traces in our simulation. So even when we close our eyes and cut off the flood of particle interactions with our retinas most of what we call spacetime simply disappears. Spacetime simply doesn't exist as a physical structure out there in a physical universe. It's entirely a data structure constructed by our simulations from particles interacting with our sense organs.

Thus Universal Reality totally revolutionizes our understanding of the universe and achieves the unification of quantum theory and relativity while simultaneously resolving all quantum paradox. Thus it's an enormously important advance that works without effectively altering either theory, both of which continue to work extraordinarily well at their respective scales. All this is achieved by just reimagining the fundamental spacetime fabric that underlies them both.

It's interesting to note that dimensional space can be said to exist only so that mass-energy can be fully conserved. If particle masses came in exact multiples of each other there would be little need to conserve some of the mass-energy of the typical particle interaction as linear velocity and as a result a dimensional spacetime might be unnecessary because particle interactions could take place without expressing energy deficits as spatial velocities. Only with the existence of particles with strange non-multiple masses do we need dimensionality because dimensionality arises only among particles. Without the peculiar particles that exist due to the complete fine-tuning there would be no spacetime.

REINTERPRETING WAVEFUNCTIONS

Though quantum theory remains essentially as it was and all its exhaustively tested equations remain in effect there is a completely new interpretation of what wavefunctions really are.

Wavefunctions are no longer particles smeared out in an exact preexisting space but rather descriptions of how inherently fuzzy dimensional fragments can align to form what we call spacetime. Wavefunctions are descriptions of the little inherently fuzzy fragments of dimensionality out of which a consistent spacetime is stitched together rather than particles smeared out in an exact preexisting spacetime.

This is clearly demonstrated by the fact that wavefunctions themselves have exact trajectories and time evolutions. This shows that the computational particles themselves are exact and only their space versus time dimensionalities are indeterminate.

Thus wavefunctions continue to correctly describe quantum events as before. However the interpretation of what they are is turned on its head.

Wavefunctions are mathematical structures representing the probabilistic dimensionality of particles. They have a real and an imaginary part. The real part is what is typically shown in illustrations and represents the time evolution of probabilities of the particle's dimensionality. It takes the form of either a linear wave along an exact trajectory or a harmonic standing wave when constrained in some manner. The imaginary part represents the probabilities of where the particle might actually be found in the wave.

So there are two components to a wavefunction, the time evolution of its state, and the probabilities of where the particle can be found in the current state at a particular time. Harmonic standing time independent wavefunctions can be more completely depicted. For example the harmonic wave of an orbital electron depicts the range of locations the electron could occupy, and the imaginary part of the orbital equation depicts the probabilities of the electron actually being found at locations within the wave. This can be depicted by the density of the wave (its transparency or opaqueness) at various locations. The greater the opaqueness the greater the probability of the particle being found in that location (Wikipedia, *Atomic orbital*).

Our processor oscillation model does exactly what wavefunctions do, but in the context of an entirely new and revolutionary new model of the universe.

THE NATURE OF PROBABILITY FUNCTIONS

Physicists tend to think of wavefunctions as actual physical entities but this is incorrect. Ultimately a wavefunction is simply a probability function rather than something that exists out there in the real world. Particles are exact computational entities with exact trajectories whose *observable* dimensionality over an exact trajectory varies probabilistically due to the oscillations in the processor application that computes them.

Thus in a decoherence there were never any other actual dimensional values than those actually chosen even though they appear randomly and weren't themselves actual until they were measured in the decoherence. The other possibilities in the wavefunction have the same status as the possibilities of a thrown dice landing on a particular side. As soon as the dice lands on one particular side all the other possibilities 'collapse' and vanish erased by the reality of the side landed. So the probability distributions described by wavefunctions were never anything real, but only a *description* of how the particle's dimensionality *might* become real. Nothing physical or real has happened to the other probabilities; it's just the description that has changed.

If a man takes the train in the morning the probability that he didn't take the train vanishes to zero but there was never anything real or actual associated with that possibility. This is the essence of wavefunction 'collapse'. The wavefunction was never something physical or even numerically real that existed, it was simply a probabilistic description of what *might* happen, and as actual events occur the descriptions of what might happen continually change with no actual consequences at all. They were never part of reality in the first place but only some observer's description of a possible future reality that never had any actual reality.

Millions upon millions of probability descriptions of practically anything can be imagined with greater or lesser accuracy. But making one up doesn't add anything to the real world, which continues on its merry way. Same with wavefunctions, they are merely descriptions of how things might become observable; they are not real in any other sense at all so there is nothing actually collapsing but the description of an unfulfilled possibility. Wavefunctions are useful in making predictions, but only in the same sense as anything else. It's not the prediction that is real but the actual behavior of what was being predicted. The whole notion that wavefunctions are somehow actual particles is entirely misguided and a wholly incorrect interpretation of quantum reality.

And this is also true about every aspect of the past including the complete fine-tuning. No matter how we can imagine any part or parts of it could have been different these are all completely imaginary *descriptions*, which never had any reality whatsoever. And every one of those possibilities has now collapsed into non-existence by the actual events of the past. Thus there never was any actual possibility of the past being even one iota different than it actually was and this includes the complete fine-tuning itself. There was never any actual possibility of the complete fine-tuning being different than it was and continues to be. All other possibilities are completely and absolutely falsified by what actually happens with no effect whatsoever on the universe.

One might claim the difference between classical and quantum predictions is that we are dealing with a complex probability amplitude in the form of a wave that can interfere with itself (as in the double slit experiment) and other predictions. But this isn't relevant. Classical level predictions are frequently interdependent. And when any prediction is itself complex with multiple interconnected aspects, changes in one aspect often affect the others. So wavefunctions are not nearly as mysterious as they are made out to be.

So wavefunctions are not real, they are not part of computational reality itself, but in our interpretation they do reveal something important about how processor cycles oscillate. They oscillate in a manner whose probabilities are accurately described by wavefunctions and the Schrödinger equation.

THE SOURCE OF QUANTUM RANDOMNESS

The elemental program computes everything exactly to the extent it has algorithms that cover them. But whenever the laws of nature have no exact decision making rules decisions must be made randomly among possible choices based on their probability distributions. This is a basic principle of reality and the source of all the randomness in the universe and the reason reality is not deterministic.

Random choices happen when events could validly emit alternate particle sets, or when particles can be emitted with alternate trajectories while still conserving energy and momentum. In such cases dimensional values are randomly chosen from available choices according to their probabilities, which generally have to do with minimizing energy levels though in the case of spin entanglement the probabilities of any equal and opposite orientation are the same because there are no energy differences.

And importantly it's also true of the space versus time oscillations of the processor that computes the observable universe. Or in the usual quantum interpretation the measurement of observables from wavefunctions where observable values are chosen randomly from their probability distributions in a process of decoherence.

It's also true of the spin orientation paradox where the decision is delayed. In the case of spin orientation the randomness occurs not as the entangled spins are created but when they are measured, as they must decohere to exact values to be conserved in events.

For example, the spin orientation of the first particle measured can have any possible random direction within 3-dimensional space. Whatever that random choice turns out to be the spin orientation of the other particle will automatically decohere to be equal and opposite since that opposite spin relationship was already part of the two particle dimensional relationship, which is now aligned to the orientation of the laboratory dimensionality.

In general quantum events are random because there aren't any mathematical rules for how to exactly align separate independent dimensional fragments, so nature must choose alignments randomly within allowable constraints. This is effected by the randomness of the processor cycle oscillations.

Most randomness at the quantum level seems to occur with respect to dimensionality as it arises computationally in the merging of separate dimensional fragments as they are aligned and created by decoherence events. Thus prior to the emergence of dimensional interactions the computations of reality can be quite simple and exact as they contain none of the complexity necessary to describe events in relationship to intrinsically fuzzy dimensionality.

Ultimately the random oscillations of processor cycles as dimensionality is computed is the source of all the randomness of the universe, since all actual randomness occurs only at the quantum level. The apparently random processes of the classical level are either quantum level randomness amplified by supporting computational structures or simply processes that are too complex to be exactly computed. Most classical level non-predictability such as weather forecasts is a combination of the two.

This mechanism of random choice due to a lack of exact decision-making rules at the quantum level is the ultimate source of all the randomness of the universe. Without this single mechanism the universe would be completely deterministic, the future would be fixed, and there would be no free will. Thus all the randomness of reality which gives meaning and frees the universe from complete determinism is an emergent manifestation of the manner in which dimensionality is computed by the processor cycles that compute the universe. It's the quantum scale oscillations of the processor that frees us and the entire universe from complete determinism and makes life meaningful.

BOUND PARTICLES

Up till now we have been concerned with free particles but of course it's particles bound in atoms and molecules that make up most of the structure of the universe.

Bound particles are essentially bound entanglements of continuous interactions, and form a major part of the entanglement network. It appears that all the complexities of atomic and molecular structures emerge naturally from the simple rules that govern the interactions of free particles. Thus atomic and molecular matter is actually an emergent phenomenon that arises naturally from inter-particle interactions according to the rules of the complete fine-tuning.

This means that all of the very complex equations that describe atomic and molecular matter are not part of the programs that actually compute the universe. These equations are *descriptions* of the aggregate behavior of the elemental program that actually computes particle events. All the emergent structure of the universe is simply the manifestation of individual particle interactions at the aggregate level.

There appear to be no higher level laws involved in actually computing the universe, though the programs of purposeful beings appear to act as such in the same manner that complex computer programs are directed towards computing various tasks but function only in terms of patterns of machine level operations. This is a complex subject that is explored further in the chapter on *Emergence*.

The atoms and molecules that make up all the mass-energy structures of the universe are composed of bound particles, in particular the electrons that occupy orbitals around the protons and neutrons of nuclei, which in turn are composed of quarks bound by the strong force carried by gluons. The rules that govern particle binding are consequences of a few simple rules that govern how different types of free particles and their particle components interact.

In atoms, electrons are attracted to the opposite charge of protons in the nucleus and drawn towards them, however they rarely have enough energy to react with a proton to produce a neutron. The mass of a neutron is significantly greater than that of a proton and electron combined. So for a reaction to occur there must be enough additional energy to be converted into the additional mass necessary to form a neutron. This can only occur with very high velocities or intense gravitational fields that provide enough intrinsic velocity energy to be converted into the necessary additional mass.

This does occur in extreme cases such as neutron stars where atoms are crushed by intense gravity and electrons do combine with protons to form neutrons and all the atoms of the star collapse to the size and density of their nuclei.

Thankfully in most cases electrons don't carry enough energy to react with protons to form neutrons, and electrons in atoms are unable to combine with protons. Thus it's only the slight mass disparity between neutrons, and protons plus electrons, that prevents all atoms from collapsing into neutrons and all the ordinary matter in the universe from disappearing! An important example of how the universe is very precisely tuned to maintain its structure.

Instead electrons are trapped in atoms by nuclear protons and continually bounce back and forth around them because they are unable to either combine with them or escape from them. The electrons become bound by the electrostatic attraction of protons and oscillate around the nucleus forming standing waves of the probability distribution of where the electron might be at any moment with respect to the nucleus.

The forms bound electron probability waves take around nuclei are called *atomic orbitals*. The basic principles underlying the forms of orbitals are fairly simple through the actual resulting forms in multi-electron atoms become quite complex due to electron-electron interactions and the imperfect spherical attraction of multi-proton nuclei (Wikipedia, *Atomic orbital*).

Because they are constrained by the attraction of nuclear protons bound electrons settle into harmonic standing waves centered on the nucleus. Harmonic waves are standing waves with an integer number of nodes that maintain their forms over time. They are analogous to a violin string fixed at both ends, which can only vibrate in a standing wave of one, two, three, or more integer nodes when plucked.

Because electrons form standing waves their energies become fixed. Each different standing wave has a specific energy and for an electron to jump between from one standing wave to another it must absorb or emit a specific amount of energy in the form of a photon. The specific frequencies of the photons emitted or absorbed accounts for the distinctive spectral colors and lines of the various elements, and the fact that atomic orbitals have discrete energies is the origin of quantum theory.

The quantized energy levels result from the relation between a particle's energy and its wavelength. For a confined particle such as an electron in an atom, the wavefunction takes the forms of standing waves. Only stationary states with energies corresponding to integral numbers of wavelengths can exist; for other states the waves interfere destructively, resulting in zero probability density.

A more accurate analogy is that of a circular drum head whose circumference is fixed to the drum rim. Depending on how it's struck it vibrates as a standing wave with one or more nodes and the waves produced in the drumhead are nearly identical in form to plain views of electron orbitals. (Wikipedia, *Atomic orbital*) has some excellent animations.

So the secret to understanding electron orbitals is they are all the possible modes of dimensional oscillations of standing waves with increasing numbers of nodes in 3-space around a center that constrains them. This is the simple key to understanding electron orbitals and underlies the periodic table of elements.

The atomic orbitals form successive *shells* of increasing radius around the nucleus. Electrons in an atom are uniquely described by 4 quantum numbers so that no two electrons in an atom can have the same 4 quantum numbers. The first quantum number n denotes the *shell* and is simply the number of nodes electron waves in that shell have; 1, 2, 3, etc.

The second quantum number l is the *azimuthal quantum number* and ranges across all integer numbers such that $0 \le l \le n-1$. Thus for each quantum number n there is a set of $l+1$ quantum numbers corresponding to the number of possible harmonic wave forms with n nodes. The azimuthal quantum number basically describes the orientations of possible standing wave forms relative to the 3 spatial axes. There are 3 possible identical harmonic waveforms, one along each axis.

The 3rd quantum number m_l, the *magnetic quantum number*, describes the magnetic moment of an electron in an arbitrary orientation and is also an integer that varies within the subshell l_0 such that $-l_0 \le m_l \le l_0$. So for example for subshell $l=2$, m_l would take on the values -2, -1, 0, 1, 2 corresponding to the possible harmonic wave forms for that n shell and l subshell. All the m's within all l's for a given n correspond to all the possible harmonic wave forms with n nodes, and together for all n's these define all possible orbitals for an atom.

The number of possible waveforms a spherically centered standing wave of n nodes can have in 3-space is n^2. However two electrons can assume the same orbital waveform if their spins are oppositely oriented so the maximum number of electrons in a shell becomes $2n^2$. *Spin*, s, is the last of the 4 quantum numbers and is always plus or minus ½ (spin up or spin down) since the electron's spin is ½.

Thus the possible orbitals electrons can occupy in an atom are simply the number of possible harmonic waveforms an electron wave can symmetrically assume centered on a nucleus.

The orbitals are the waveforms that actual electrons must assume when they are filled. The actual number of electrons in a neutral atom is equal to the number of protons so the charges are balanced, and the number of protons determines the element in the periodic table.

In turn atoms form molecules primarily by sharing outer orbital electrons. When atoms combine in molecules outer atomic orbitals become distorted into molecular orbitals and the specifics of how these form and their complex properties determine the laws of chemistry and thus the structure and interaction rules of all matter. Thus chemistry and all mass-energy structures are the emergent results of the bound interactions of electrons, protons and neutrons.

A somewhat analogous situation occurs in the nucleus where the mutual repulsion of protons is overcome by the strong force, and proton and neutron quark waves vibrate in complex standing wave forms around each other. However, because the masses of protons and neutrons are much greater than the electron mass their de Broglie waves are much smaller, and thus nuclei are much smaller than electron orbitals.

The orbital forms shown in most illustrations are those of ideal individual harmonic waves. They are those a single electron would assume around a single proton as it increased or decreased its energy and jumped from orbital to orbital. However the presence of multiple electrons occupying orbitals in a single atom distorts their orbital forms due to the mutual repulsion of electrons and the necessary presence of multiple protons so the positive charge of the nucleus is not exactly spherically symmetric.

Thus though the principles underlying electron orbitals are fairly simple the forms they actually take in atoms become quite complex with increasing atomic number due to the mutual repulsion among electrons, including the screening of positive nuclear charges from electrons in larger shells, the uneven attraction of multiple protons in the nucleus, and even relativistic effects coming into play in larger orbitals.

In more technical terms the electron wavefunction oscillates around the nucleus according to a time independent (unchanging) Schrödinger wave equation, and orbitals are its standing waves. The standing wave frequency is proportional to the orbital's kinetic energy. The real part of the Schrödinger equation gives the form of the orbital, and the imaginary part gives the probability distribution of finding an electron at a particular location within it.

Thus emergence begins at the atomic level more or less. Atomic and molecular structures emerge automatically from the way that electrons, protons and neutrons interact, from the computational rules that actually compute their interactions. Thus individual particle interactions are all the elemental program needs to actually compute, and the computational universe becomes very much simpler than we might expect.

Electron orbitals, and thus all chemistry and the structure of all matter emerges from the possible harmonic forms bound electron waves can take around protons, and electron waves themselves are the result of the space time oscillations in the processor cycles that compute reality.

Thus a small set of fairly simple particle interaction rules produces the atoms and molecules that compose all material structures:

1. The mutual attraction of electrons and protons.
2. The inability of electrons to actually combine with protons at normal energies.
3. The resulting possible orbital forms of standing harmonic electron waves centered on the nucleus.
4. The analogous strong force binding of proton and neutron quarks in nuclei.

Thus the structure of the observable universe from compound particles and atoms on up is emergent rather than independently computed. All the actual computations of reality occur at the elementary particle and particle component level, and larger scale structures automatically emerge from bound particles that manifest these elemental computations. This is how all the incredible complexity of the observable universe emerges from the finely tuned interplay of the simple elemental rules that actually compute it. This greatly simplifies the computational structure of reality. It's the incredibly amazing complete fine-tuning that is responsible for the wonderfully meaningful complexity of emergent structures that are all the aggregate manifestations of computations of individual particle interactions.

Because they exist as bound entanglements the dimensional relationships of electrons and protons are exact in terms of energy conservation, which is of course the basis of quantum theory. However as in other cases when these exactly conserved energy relationships are viewed in terms of dimensional positions and velocities of electrons relative to the nucleus they are subject to the processor cycle oscillations as viewed from an observer frame, which is why they appear as harmonic standing wavefunctions within which the actual position and velocity of the electron appears probabilistic.

TIME

THERE ARE TWO KINDS OF TIME

One of the most important insights of the theory of Universal Reality is a completely new understanding of time. The most fundamental personal and scientific observation of our existence is a present moment through which clock time flows. We now have an elegant computational mechanism that explains both these self-evident aspects of time.

First it's the *presence of reality* that manifests as the present moment. To be real the universe must have presence and this presence is what we experience as the present moment through which clock time flows. Because the presence of reality is universal there is a single universal present moment within which the entire universe exists.

Now because happening is an intrinsic characteristic of reality, time flows through the present moment. The time of the present moment is not static but flows as the entire data state of the observable universe is continually recomputed in the current present moment. This can be called the flow of present moment time or *P-time* for short.

Now the flow of P-time continually drives the computations of the universal program. It drives the universal processor that continually recomputes the data state of the entire universe.

The fabric of the universe consists of a single ubiquitous field of spacetime velocity in computational space that has the value of the speed of light at every point. This uniform c valued velocity can take the form of either velocity in space or velocity in time. To the extent it manifests as some form of spatial velocity the velocity in time is reduced so that the vector sum of both velocities continues to equal the speed of light.

The remaining amount of c valued spacetime velocity that takes the form of velocity through time manifests as the local clock time rate. It's our local clock time rate that we experience as the flow of clock time within the present moment. Clock times flow at different rates depending on the amount of spatial velocity present but we always experience clock time having the same *apparent rate* because our mental processes are slowed by the same amount.

This is the explanation of our fundamental experience of a universal present moment through which clock time flows at different relativistic rates.

Thus there are *two separate kinds of time*. First there is the universal time of the present moment that flows at the same rate throughout the universe. Its source is the fixed cycle rate of the processor that continually recomputes the entire universe in the current present moment including all local clock time rates. Within the constant flow of P-time the processor computes the rates of all local clock times based on the spatial velocity present.

There is simply no way around this fundamental fact and we directly experience and confirm it at every second of our existence as a present moment through which clock time flows, and through which the clock times of all processes flow at rates depending on their spatial velocities.

This is also confirmed by relativity itself because no matter how local clock time rates differ they all do so in the same universal present moment common to all observers. When space travelers return to earth and compare clocks with stay at home observers they always do so in the same common present moment. If they weren't in the same common present moment it would be impossible for them to compare their relativistic clock times. Thus relativity itself actually requires a separate present moment time to even work.

Unfortunately many physicists don't understand this because they are unable to comprehend the self-evident existence of a present moment. A number of convincing arguments demonstrating two kinds of time are detailed in my other books in particular *Understanding Time, What it is and How it Works* (Owen, 2016). We can also offer a couple of simple proofs here.

PROOFS OF TWO KINDS OF TIME

We can conclusively prove there are two kinds of time from relativity theory itself. All that's needed is first to show that every proper time reading of any clock always occurs in the current present moment of that clock, and second to prove there is a single unique set of proper times for all the clocks in the universe at any time on any clock. This will prove that the current present moment times of all the clocks in the universe have a single unique one-to-one relationship, and therefore there must be a single universal present moment time common to all the clocks in the universe.

Here is the logical common sense proof:

1. Every clock always ticks in the current present moment it experiences. Therefore the passage of time on any clock is its record of the passage of present moment time.
2. For every tick on any clock in the universe all the other clocks in the universe must also be ticking some time in *their* experience of their current present moment.
3. Therefore all the clocks in the universe are always simultaneously experiencing a current present moment as they tick.
4. Whenever clocks meet in the same location we find their current present moments are the same current present moment. Therefore there must be a single universal current present moment time common to all clocks.
5. No matter what different times clocks read or how their rates vary across the universe they all continually tick their proper times in the same universal current present moment.
6. Thus there is a single universal present moment time across the entire universe common to all observers and all processes within which local clock times flow at different relativistic rates depending on the amount of spatial velocity present.

A more formal proof based on the invariance of proper time (http://www.astro.ucla.edu/~wright/relatvty.htm) can also be given.

First a couple of standard concepts from relativity are needed. *Proper time* is the time an observer reads on his own clock. And a *proper time interval* is the time a clock takes to move from one event to another along a world line as read on that clock. Proper time is the elapsed time on a clock moving from one event to another. *Coordinate time* is the time an observer reads on a clock *that isn't his*. A *coordinate time interval* is the elapsed time it takes on a second observer's clock for a proper time clock to move between two events.

1. A basic principle of relativity is that *proper times are invariant*. This means that the time a clock reads as it participates in an event is seen the same by all observers no matter what their own

frames are. If a clock reads 10:01 at an event it participates in then every observer in the universe agrees it reads 10:01 at that event. This is just basic common sense but it's confirmed by relativity.

2. Now consider a clock that participates in a series of events. Each of these events is the person holding the clock reading its proper time.

3. Each of these readings takes place in the current present moment time experienced by that person.

4. Now assume I am watching that person read his clock as I move relative to him far away in a different gravitational field. By the invariance of proper time I still see him getting the same readings of proper time that he does. Our clock times are now flowing at different rates and we have different proper times on our own clocks but still I agree with his proper time readings on his own clock.

5. Now every observation of him reading his clock takes place at some proper time on my own clock and thus occurs in my experience of the current present moment of time.

6. Thus there is always a unique one-to-one correspondence between his proper time and my proper time. And because the invariance of proper time is true in any direction we both agree what this correspondence is. We both agree what proper time reading on his clock syncs with what proper time reading on my clock, and this is true for every reading on either of our clocks. There is always a single unique one-to-one correspondence of every one of his proper times with one of my proper times and vice versa no matter what those times are and how fast our clocks are running relative to each other.

7. Now I can always determine what *specific* proper time of mine corresponds to any proper time of his by simply subtracting the time it took for light to travel from it to me from my current proper time. And he can do the same.

8. Assume his proper time p1 occurs in the same current present moment as my proper time p2. I see his p1 event when my proper time reads p2 + Δc where Δc is the time it took light to reach me from his p1 on my clock. So to determine what proper time of mine corresponds to his p1 I merely subtract Δc from my current proper time and arrive back at p2. And he does exactly the same to arrive back at p1 on his clock. Thus we both arrive back at the same proper time correspondence p1 = p2, and both confirm that his p1 corresponds to my p2. Even though he and I may measure Δc differently if our clocks are running at different rates we both arrive back at the correct proper time correspondence because we both subtract exactly what was added.

9. Now the invariance of proper times holds across all observer frames. Thus there is a single unique one-to-one correspondence between the proper times of all the clocks in the universe that is the same for all observers in the universe. If my p2 = his p1 and there is another observer whose p3 = his p1 then my p2 = p3. Thus there is a single unique one-to-one correspondence among all the proper times of all the clocks in the universe for each proper time on any clock in the universe.
10. Thus for every reading on any proper time clock there is a single internally consistent set of proper time readings with all other clocks in the universe that is the same for all observers.
11. The proper time reading of every observer's own clock always occurs in his experience of the current present moment.
12. Since there is always a single unique proper time on all the other clocks in the universe corresponding to every proper time reading of any observer's own clock, the present moment experiences of all observers in the universe are also synchronized.
13. This proves that there is a single universal current present moment time across the entire observable universe common to all observers.
14. Thus all observers in the observable universe exist in the same current present moment through which their own proper times flow at local relativistic rates depending on their spatial velocities. All local clock time rates in the observable universe run within a single common universal flow of present moment time.

The common sense interpretation is that at any time on my own clock something must be happening everywhere else in the universe. And whenever anything happens it happens in the present moment. Therefore if something is always happening everywhere in the universe it must be happening in the current present moment, and that current present moment time must be universal and common to all the happening in the entire universe. Thus there must be a universal present moment time that flows at the same rate throughout the entire universe within which all local clock times are computed.

More fundamentally the existence of two kinds of time is experimentally confirmed by relativity itself. Every time astronauts return to earth their clocks read slightly different times than earth clocks. But before, throughout, and after their missions they always remain in the same current present moments we experience here back on earth. It's only in a common present moment that different times running at different rates can even be compared. Thus relativity itself absolutely requires a universal common present moment to even make sense.

The very fact that clocks that ran at different relativistic rates are compared back here on earth in a common present moment conclusively demonstrates there are two kinds of time. It's also consistent with the STc Principle that underlies relativity and all the additional insights that emerge from it. Thus there is absolutely no doubt at all of the existence of two kinds of time. This realization opens the way to all sorts of amazing new insights about the very nature of reality and how the universe operates.

Even so many physicists still mistakenly reject the very notion of a present moment based on an out of date interpretation of relativity. Instead they tend to believe in the positively delusional *block universe theory* in which all times exist 'at once' and nothing actually happens. This theory is intrinsically inconsistent and easily falsified as in *Understanding Time, What it is and How it Works* (Owen, 2016).

It's worth noting that every proper time clock runs at the local speed of light because observers have no linear spatial velocity relative to their own proper time clocks. Thus every observer continually travels through time at the speed of light at every moment of his existence. We are continually traveling through time at the speed of light in every moment of our existence.

P-TIME

If we call the time of the present moment P-time, then every actually occurring event takes place simultaneously for all observers at the same P-time throughout the entire universe. In other words if an event takes place in the present moment for one observer, it also takes place at the exact same P-time for all observers. This will always be true no matter how different the clock times and clock rates of various relativistic observers are. This universal simultaneity of P-time is what allows relativistic observers to compare their different clock times in the same present moment.

This P-time simultaneity has nothing to do with the *relativity of simultaneity of clock time*, which is a well-established consequence of relativity (Wikipedia, *Relativity of simultaneity*). Different observers can still have different views of the clock time simultaneity of events due to the finite speed of light between observers and events. However the entire observable universe, including its different clock times, is continually recomputed in the same universal P-time moment.

In many cases this proper time correspondence can be calculated and confirmed. If the space traveling twins exchange flight plans before they separate the one-to-one correspondence of their proper times throughout their separation can be calculated. Each will know the complete relativistic history of the other and thus know exactly how much proper time has elapsed on the other's clock for any proper time on his own clock. Each will know what the other's clock is reading at every moment on his own clock. To be absolutely clear this *calculation* of the current *proper time* of the other twin is not the *coordinate time* that would be *observed* on the other twin's clock. It's the *proper time* of the other twin's clock, which is not generally observable but which can be calculated.

Thus it's always possible for any observer who has knowledge of the relativistic circumstances of any other observer to calculate what proper time reading of that other observer correlates to each of his own proper time readings. There is always a one-to-one correspondence of proper time readings between any two observers that tells them what proper time of one corresponded to the proper time of the other in their common present moment even when they are separated in different relativistic circumstances with different clock time rates.

However in the general case of any two observers in the universe, where an initial proper time correspondence can't be determined or the relativistic history of the other observer is not known, it may be impossible to calculate the current proper time correspondence even though it's certain one must exist. However if any observer in the universe can determine the relativistic variables of any other observer as well as his own he can calculate the proper time rate of that observer relative to his own proper time rate and confirm the existence of a common shared P-time.

Thus it's easy to show there will always be a one-to-one proper time correspondence between any two observers in the universe, and this is all that's necessary to conclusively demonstrate a universal P-time present moment common to all observers. It's sufficient to note that time is continuous for all observers thus every observer in the universe must be doing something at every proper time moment of any other observer's clock. Whatever is being done will be done in the exact same common universal present moment of all existence in the universe because the present moment is the only time that anything can occur because it's the only moment that exists and the only locus of reality.

The clock times of different observers can flow at different rates through this common present moment, but there is always a one-to-one correspondence between the proper times of any two observers throughout the entire universe. This is consistent with Universal Reality's proposal that the common universal present moment of existence is all that exists, and is the current moment of happening in which all the computations of the universe take place. It is the current universal P-time tick of the entire computational universe.

P-time has no intrinsic metric of its own since it's the computational source of all clock time metrics. Therefore the proper times of observers must be used to notate the passage of P-time, and their correspondences, when they can be determined, can be used to establish identical P-times among observers. And if all else fails any two observers can simply communicate their current relativistic conditions and proper times to enable their P-time simultaneity to be calculated.

ACTUAL VERSUS OBSERVATIONAL RELATIVITY

We can define two distinct types of relativistic effects, *actual* and *observational*. Observational effects vanish as soon as the relative motion between observers ceases, while actual relativistic effects have permanent effects that all observers agree upon when relative motion ceases.

For example when two clocks move relative to each other they each observe the other clock run slower due to its apparent spatial velocity. This is an *observational relativistic effect* and the two observers don't agree on whose clock is actually running slower.

However when the relative motion stops, only the clock that has actually moved relative to the absolute computational frame of the entanglement network shows less elapsed time and at this point both observers will agree to this. This is an *actual relativistic effect*.

This can only be explained if there is a universal spacetime metric relative to which actual relativistic effects occur, and this can only be the entanglement network in which everything is actually computed.

This model also immediately solves the problem of what *rotation* is relative to. Whether something is actually rotating or not doesn't depend on the motion of any observer. It's absolute and reveals itself by the presence of centrifugal force. It's either rotating or it isn't irrespective of the motion of any observer. It always shows itself in whether or not water climbs the sides of a rotating bucket or whether a gyroscope is resistant to tilting or not.

What rotation is relative to has been an unsolved problem for at least several hundred years since Newton first described his problem of 'Newton's bucket' (Wikipedia, *Bucket argument*). Over a hundred years ago Ernst Mach proposed that actual rotation was with respect to the aggregate mass of the universe but neither relativity nor any other theory has been able to provide any explanation of why this might be true (Wikipedia, *Mach's principle*).

However in a computational universe the answer is clearly that actual rotation is with respect to the actual spacetime background in which it's computed, which is in fact roughly aligned with the aggregate mass distribution of the universe. This is a direct consequence of spacetime being computed by particle interactions. Thus there must be an actual universal spacetime metric to explain the absolute nature of rotation, and an actual universal spacetime metric relative to which motion through space produces actual as opposed to observational relativistic effects.

But there is even stronger evidence for a single universal spacetime. For relativity to even make sense there must be a single universal present moment for relativistic clock times and motions to be compared. If there wasn't a common present moment to compare them then different clock times couldn't even be compared and relativity would become nonsensical!

Thus the existence of an actual universal background spacetime metric is implicit but largely unrecognized in relativity itself. It's absolutely necessary for relativity to even work and make sense. A common present moment is absolutely necessary for different relativistic frames and clock times to exist within and be compared.

A universal present moment is also a direct consequence of relativity in another way. The STc Principle tells us that everything continually moves through combined space and time at the speed of light.

Thus everything must necessarily always be at a single point in time. That single point in time is quite obviously the present moment.

Thus contrary to what most physicists believe, relativity itself requires the existence of a common universal present moment, and in a computational reality that present moment is the universal P-time in which the data state of the entire universe is continually recomputed.

This absolute computational spacetime metric is the dimensional consistency of the entanglement network built up from all particle interactions in the universe. Since all subsequent dimensionality is computed with respect to the consistency of this dimensionality actual versus observational relativistic effects occur when there is motion relative to it. So there is effectively an absolute background spacetime but it's a computational structure in computational space rather than a 4-dimensional physical structure.

THE UNIVERSAL SPACETIME METRIC

The equations of relativity are able to treat all frames identically because observational relativistic effects obey the same transient rules as actual relativistic effects do. This is because relative motion and actual motion relative to the computational background appear the same observationally. Only rotation and the persistent mutually agreed effects of actual spatial velocity relative to the computational background give the game away.

The Einstein field equation is the fundamental equation of general relativity. In the field equation the presence of mass-energy (the stress-energy tensor) determines the Einstein or *metric tensor* that describes the form of spacetime resulting from the presence of mass-energy at any point. As popularly stated the field equation describes how 'the presence of mass tells spacetime how to curve'. The Einstein field equation is described in simple detail in *Relativity Made Easy* (Owen, 2016).

Now the field equation describes the curvature of space *at any point* due to the presence of mass-energy at that point. And it does this in terms of generalized coordinates that can be mapped to the actual coordinate frame of any observer. So by considering all points the field equation allows any observer to potentially derive a complete metric that accurately describes the curvature of all of spacetime from the perspective of his location and motion within it. The field equation does this beautifully and over more than 100 years and likely millions of calculations it's always proven correct.

However the field equation isn't the whole story because it fails to explain the apparent *absolute nature* of certain aspects of spacetime such as acceleration and what can be called *actual* versus *observational* relativistic effects. Specifically general relativity can't explain

1. What rotational and linear *acceleration* are with respect to, why some motions produce the intrinsic feelings and scale deflections that identify acceleration and some don't. Physicists may correctly state that acceleration is motion with respect to a local inertial frame but that's simply the definition of a local inertial frame rather than an explanation of why some frames are inertial and some aren't. It misses the problem.
2. Why some spatial motion produces *permanent relativistic time dilation* and some doesn't. For example both twins each move relative to the other at the same speeds over the same distances yet when they meet only the twin that traveled away from earth and returned has aged less and shows less elapsed time on his clock. What was he traveling with respect to that explains this? General relativity doesn't explain it. It fudges the answer by assuming a spacetime metric that gives the right answer but never explains the source of that metric or what it actually is.

These two effects can only be properly explained on the basis of motion with respect to a *single actual universal spacetime metric* that is properly outside the scope of the Einstein field equation itself and can only be accommodated by tinkering with its meaning and assuming frames that give the correct answers without explaining their source or actuality.

The field equation itself neither predicts nor describes a single actual universal spacetime metric. But a single actual universal spacetime metric must exist to completely explain actual relativistic effects. Only the computational model of Universal Reality provides an adequate explanation of these effects by showing how they depend on the existence of an actual universal frame independent computational spacetime metric and explaining what this metric is and how it's created.

General relativity describes spacetime exclusively in terms of observer-based views but fails to recognize there must be a single actual universal spacetime metric common to all of them that observer views are views of. There must be a single actual universal spacetime metric to explain *acceleration* and *actual permanent relativistic effects*.

Physicists manage to sweep these problems under the rug by constructing generalized models of spacetime based on highly generalized views of observers and mass-energy distributions, but the fundamental problems still remain though they are typically unrecognized or even actively denied.

Even though many physicists deny it there simply must be a single universal spacetime metric to explain all observational facts. In fact the greatest physicists of the past such as Newton, Mach, and Einstein did recognize this problem in their writings but were unable to solve the source or nature of this universal spacetime metric.

Universal Reality provides a simple straightforward solution to this problem by describing what this universal spacetime metric is, how it's computed, and how it solves these fundamental problems. It does this by placing general relativity and quantum theory in the context of a universal Theory of Everything within which they both continue to operate exactly as before. Thus this Theory of Everything is completely consistent with both theories but places them both within a more fundamental and complete unified model of reality.

The dimensional aspects of the previously described *entanglement network* are in fact the necessary *actual universal spacetime metric*. The Einstein field equation describes spacetime exclusively in terms of observer views all of which it considers equally valid. However it completely fails to specify what the actual universal spacetime metric these views are views of actually is.

In contrast Universal Reality says there must be an actual universal spacetime metric computed by all particle interactions that takes the form of the resulting dimensional interrelationships of particles. In Universal Reality everything is the data of what it is and fundamentally everything is the data of elementary particles and their particle components. This is the actual stored data of the universe that constitutes the observable universe.

So the Einstein field equation accurately describes observer views of this structure but it has nothing to do with computing it whatsoever. The field equation *describes* but doesn't *compute* the relativistic aspects of the observable universe. And it only describes observer views of the actual universal spacetime metric that underlies them all and which provides the explanations missing from general relativity itself. It's impossible for the field equation itself to actually compute an actual universal spacetime metric because it's frame based and it would be impossible to store the infinite actual data of the infinite number of possible frames within the universe. Nothing would be up to that job.

Thus rotation and actual relativistic effects are due to motion relative to an actual underlying universal metric that is completely missing from general relativity itself. However motion with respect to the universal metric itself must be carefully and correctly understood:

1. The existence of a single actual universal spacetime metric can be confirmed, detected and measured by acceleration (linear or rotational) because acceleration is a change of velocity with respect to the metric itself rather than with respect to other objects. This is why proper acceleration (the acceleration an object actually feels) is invariant. So whenever acceleration is felt it's due to a change in spatial velocity with respect to the universal spacetime metric and reveals the metric by its presence.

2. Non-accelerated (inertial) spatial motion (velocity) with respect to the metric can be determined by the resulting *permanent time dilation* (the slowing of time velocity). However permanent time dilation can only be observed by comparing clocks. It can't be directly measured because clocks slow at the same rate time slows so proper time always appears to run at the same rate. However when for example the twins meet up they can compare the elapsed time on their clocks and both agree that the traveling twin's clock shows less elapsed time so he must have had more spatial motion with respect to the spacetime metric. This enables the relative amount of spatial motion to be determined.

3. Likewise two clocks in different gravitational fields can be compared to confirm the clock in the stronger field is running slower because gravitational fields are areas of intrinsic spatial velocity that slow the velocity of time in the field. But again the observer in the stronger field can't directly observe his clock is running slower because both time and his clock are both slowed by the same amount.

4. Thus an observer can't directly measure how much of his own inertial (non-accelerated) motion is with respect to the metric because his means of measurement are affected the same as what he's trying to measure. Only by comparing his observations with others can he determine the truth. The metric itself is invisible and so can't be used as a visual reference for motion relative to it.

5. Thus with the exceptions noted motion can only be observed with respect to other *objects* in the metric, rather than with respect to the metric itself.
6. Nevertheless motion (spatial velocity) with respect to the universal spacetime metric does produce actual reduction of velocity in time according to the STc Principle that is confirmable and agreed by observers by comparing their observations.

So basically the metric itself is invisible because it's simply the emergent logico-mathematical dimensional consistency of the entanglement network. Thus unchanging (inertial) motion can only be *directly* measured with respect to other objects, but can be *indirectly measured* by comparing actual permanent relativistic effects. Whereas changes in motion (accelerations) directly reveal the presence of the spacetime metric and measure changes in motion with respect to it.

How the entanglement network is computed in a manner that unifies relativity and quantum theory was explained in the chapter on Quantum Reality. But how a single universal spacetime metric, with respect to which actual versus observational motion occurs, emerges from it requires some additional explanation.

1. At its most fundamental level assume the universe is a computational space prior to any dimensional spacetime. Computational space is analogous to computer memory which is linear rather than dimensional but within which various pre-dimensional structures such as data arrays can be defined.
2. The data that constitutes the entire universe consists entirely of current particle component data values in computational space in the form of the data of elementary particles.
3. This data that constitutes the entire observable universe is continually recomputed in every universal P-time tick.
4. These computations take the form of particle interactions in which the data values representing mass-energy are conserved.
5. As a result of the conservation of mass-energy data the particle component values of resulting particles bcome interrelated. In particular, with respect to the universal spacetime metric, the dimensional particle component values of resulting particles are interrelated.
6. Through the universal network of interactions dating back to the big bang the dimensionalities of all particles in the universe become interrelated in a single universal network.

7. The result is a single *universal entanglement network* whose internal dimensional consistency is the computational basis of the universal spacetime metric.
8. The entanglement network is constructed locally from individual particle interactions and takes the form of the resulting dimensional particle component values including numeric positions and velocities of resulting particles. The current present moment values of this data are the only actual existent stored values of the observable universe. These values are the only data that constitutes the entire observable universe.
9. The internal dimensional consistency of the current data state of the entanglement network is the universal spacetime metric with respect to which actual motion and actual relativistic effects occur.
10. However the entanglement network is constructed locally and probabilistically at the individual particle interaction level (see the discussion in the chapter on Quantum Reality) so its dimensional consistency across the entire metric is not perfectly exact due to computational lags across the network. (See the upcoming chapter on Dimensional Drift for a discussion.)
11. However the results of myriads of successive particle interactions of different particles in the network gradually converge the relativistic scale dimensionality to a single consistent spacetime metric roughly isomorphic to the aggregate distribution of mass-energy in the universe at least on local scales.
12. The resulting spacetime metric converges to the local distribution of mass-energy precisely because it's the interactions of local particles that create its dimensionality. And this consistency presumably extends throughout the universe though perhaps less precisely which is why acceleration and actual motion with respect to the spacetime metric is roughly equivalent to motion with respect to the aggregate mass of the universe as Mach suggested. Every particle interaction tends to average the resulting dimensionality of the resulting particles relative to that of preceding particles. In this manner continuing particle interactions tend to converge the aggregate dimensionality produced on that of the aggregate average of particle motions and distributions.
13. This is because all new particle interactions take place in the context of the existing dimensional consistency of the local entanglement network so that new motions are always computed with respect to it with limits of precision noted in the chapter on Quantum Reality. *The way dimensionality is computed to produce the metric is precisely the same reason that subsequent particle dimensionalities are computed with respect to it*. In this manner

the dimensionality of the metric emergently converges on the actual distribution of mass-energy. Thus the dimensionalities of new particle interactions are computed with respect to the average velocity, rotation and gravitation of the average distribution of mass-energy. This is the source of the universal spacetime metric's emergence from the entanglement network. Because all dimensional relationships are actually computed with respect to the entanglement network they tend to perpetuate the consistency of the spacetime metric, which is a generalization of the internal consistency of its individual particle relationships.

14. This universal spacetime metric is the previously discussed universal c valued spacetime velocity 'fabric' of the observable universe. As described in the chapter on Relativity & Spacetime, any spatial motion with respect to the logico-mathematical 'fabric' of the metric reduces the associated velocity of time so their vector sums always remain equal to c. This fundamental STc Principle underlies most of relativity.

This explains how an actual universal spacetime metric is created, what it is and why motion with respect to it constitutes actual motion that produces actual permanent relativistic effects. Actual permanent relativistic effects are permanent because they are computed relative to it, and remain in the data, whereas observational relativistic effects vanish as soon as relative motion stops because they were computed relative to other objects rather than to the metric itself.

It's a simple straightforward theory that explains what's missing from general relativity. And it also simultaneously unifies relativity and quantum theory, and resolves all the apparent paradoxes of quantum phenomena. So there is plenty of good evidence for this part of Universal Reality.

In some of my previous writings I've described this universal spacetime metric as an absolute computational *frame*, but this is misleading, as a frame is properly a *point-centered observer view* of the universal spacetime metric. Nevertheless the universal spacetime metric is the actual computational data structure of the observable universe. It's completely consistent with general relativity and explains the actual relativistic phenomena that general relativity simply can't. And in conjunction with the field equation it correctly describes the way the mass-energy and spacetime universe works from the perspective of all possible frames within it.

It's very important to understand that the entanglement network itself consists of the dimensional relationships among all particles on an individual particle basis. Thus the metric itself is an *emergent logico-mathematical structure* representing the underlying consistency among all the dimensional relationships in the entanglement network. It's not an actual physical structure. It's simply a conceptual map created from these individual relationships analogous to how any 3D graph is constructed to encompass some data set and then the data is displayed as points in the graph designed to contain it.

But spacetime like the map didn't exist before the points. It was constructed on the basis of the data points to display them together visually to make better conceptual sense of them. The human simulation of a 'physical' spacetime is exactly that, a map our minds create from the dimensional consistency of relationships among objects that it then places those objects within. In this manner our simulation of dimensional relationships among objects is mapped onto a conceptual map we call physical spacetime so as to make better sense of those relationships.

The final step in this process is our simulation reifies this conceptual map, projects it out around us, and convinces us this map is a physical spacetime when actually everything we see around us exists entirely as neural data within our own brains.

So there are actually *four levels* to what we think of as spacetime:

1. **Computational space**. Everything in the universe exists as *numeric* particle component data in a non-dimensional data structure similar to that of computer memory. All the actual computations of the universe occur in this computational space where all the data of all particle components is continually recomputed in every universal P-time tick.
2. **The entanglement network**. The *emergent network* of consistent dimensional relationships produced by successive interactions of all particles in the universe over the history of the universe. The current present moment time slice of the entanglement network is the emergent level structure of the actual observable universe. It's the view that emerges from considering all current particle component values in aggregate.
3. **The universal spacetime metric**. This is an encompassing conceptual map or graph extrapolated from all the dimensional relationships of the entanglement network. All the dimensional relationships of the entanglement network can be conceptually

placed within this map in a consistent manner to make better sense of their overall context and interrelationships.

4. **'Physical' spacetime**. Our mind's reified view of the universal spacetime data metric, which our mind simulates to us as the apparent physical spacetime we seem to experience ourselves and other objects within. This is analogous to how the numeric data of a virtual reality in a computer memory can be projected into the semblance of a moving 4D world in a virtual reality headset.

Because all these levels are approximately inter-consistent we can consistently make sense of things at any level. Relativity more or less describes the universe in terms of the universal spacetime metric. As humans we function within the world as if it existed in a physical spacetime. Quantum processes more or less operate at the level of the entanglement network, but everything is actually computed at the fundamental level of computational space. Universal Reality describes this entire system as one unified structure.

It's self-evident that for an actual universe like ours to be produced either computationally or physically that there must then be a single universal metric that describes it. But general relativity's field equation doesn't actually recognize that because it explains everything in terms of individual observer views of a metric but ignores the source and nature of the metric itself. But there must self evidently be some single actual universal structure to have frame views of outside the scope of the field equation. If there weren't a single actual metric the frame views were views of, the field equation simply wouldn't be able to express them consistently in a single covariant equation.

And amazingly, and without most physicists realizing it, relativity itself implies Universal Reality's theory that it's particle interactions that actually compute the spacetime metric. This is because the Einstein field equation states that it's the presence of mass-energy that 'curves' spacetime and of course it's well known that it's particle interactions that result in the presence and distribution of mass-energy.

So at the most fundamental level general relativity confirms a major theory of Universal Reality without physicists apparently even realizing it! The Einstein field equation that's the core equation of general relativity implies this but in itself only shows how any possible observer would observe its dimensionality.

So the Einstein field equation implies that the actual distribution of mass creates a spacetime metric but actually describes only individual frame views of it and so is unable to tell us what the actual universal spacetime metric actually is or where it comes from and thus misses an essential aspect of how it works to produce actual relativistic effects.

This Universal Reality model of the universal spacetime metric resolves this and appears to be the only possible way to explain all of general relativity, though the exact details of how the universal spacetime metric emerges from particle interactions need to be confirmed and further explored via computer simulations.

In a broader sense there is no actual motion with respect to the fundamental medium of *computational space and P-time* because that's the ultimate computational source of the spacetime metric with respect to which motion occurs. So all motion is computed with respect to the dimensional structure (the 'fabric') of the universal 4-dimensional spacetime metric. Any motion in *space* relative to the metric results in less motion in *time* so the vector sum of their velocities is always equal to c. Properly speaking there is motion *within* the metric which always expresses as some form of mass-energy which always slows the velocity of clock time, but *no motion relative to* the entire 4D spacetime metric itself since everything exists within it.

DIMENSIONAL DRIFT

Because dimensional spacetime is computed by successive particle interactions all current particle interactions are computed with respect to the past network of the entanglement network. All the nodes and links extending into the past from any event established a dimensional consistency that is at least partially used to create the dimensionality of the particles exiting the new event.

However as we've seen this dimensional consistency is not exact at the particle level. Locally at the classical level it is exact and takes the form of an absolute consistent network of dimensional relationships that particulate objects will have consistent dimensional relationships to.

However because the universal spacetime metric is computed locally by particle interactions there is a possibility that its overall dimensional consistency across the universe could vary from location to location over time. Thus for example spatial velocities and rotation orientations with respect to the background might appear slightly different in different locations of the universe, and also in different eras of the universe since the background has been progressively computed over time.

These possible variations in the consistency of the entanglement network across space and time can be called *dimensional drift*. Dimensional drift could show up as slight anomalies in relativistic measurements over large distances or between different time periods. It's likely that single measurements of dimensional drift would simply be mistaken as accurate descriptions relative to dimensionality back here on earth. However comparisons between several such measurements might produce slight inconsistencies that revealed actual dimensional drift.

Thus if otherwise unexplained anomalies in relativity are discovered they could possibly be due to dimensional drift. If so this would be an important test and confirmation of Universal Reality. In fact at least one such anomaly is known that doesn't appear to be completely explained that could possibly be an effect of dimensional drift (Wikipedia, *Pioneer anomaly*).

THE ARROW OF TIME

The STc Principle neatly solves the problem of the source of the arrow of time. Because everything is continually moving through combined space and time at the speed of light everything must continually move through time in a single direction.

Thus the STc Principle puts the arrow of time on a firm physical basis as a consequence of relativity itself. This is an important solution to a problem that has long troubled science.

So contrary to what many physicists believe the arrow of time has nothing to do with *entropy* for two obvious reasons. First entropy varies widely throughout the universe but that variation has no correlation with different rates of clock time.

Second, entropy isn't fundamental because entropy states depend on the strength and mix of prevailing forces. If for example gravity were simply converted to the spatial velocity of heat as it might have been in the big bang entropy states would reverse as well. (See the upcoming chapter on *Cosmology* for a discussion).

In a static universe with only attractive gravity the maximum entropy state is a universal black hole, but if all that gravitation was converted to heat then the maximum entropy state of the black hole automatically becomes the minimal entropy state of the big bang. If gravity were converted to heat at the big bang and during inflation that would neatly solve the problem of why the initial entropy state of the universe appears to have been the most unlikely minimum state. Penrose and others have unsuccessfully struggled with this problem (Penrose, 2005).

Thus the eventual maximum entropy state of the universe depends on the ongoing mix and distribution of the four forces and its current spatial expansion rate. Thus entropy is not fundamental and cannot be the source of the arrow of time. The source of the arrow of time is clearly the STc Principle underlying relativity and so much else.

THE UNIVERSE IS A HYPERSPHERE

Without an understanding of present moment time and no single universally consistent dimension of clock time cosmologists have been unable to come up with a consistent 4-dimensional geometry for the universe. Clock time clearly doesn't work as a universal 4th time dimension because there is no single universal clock time dimension valid for all observers. Thus there is no consensus among scientists as to what the geometry of the universe actually looks like at its largest scale.

However the recognition of a single universal present moment P-time immediately solves this problem. The geometry of the universe is now clearly a simple hypersphere with the surface the 3-dimensions of space and P-time its radial dimension from the surface back to the big bang at the center. Without understanding there are two kinds of time this model is impossible and thus cosmology's current model of the shape of the universe is a clearly untenable expanding trumpet (Wikipedia, *Shape of the universe*).

The universe is a finite closed hypersphere with no edges. It clearly can't be infinite, as many scientists believe, as nothing actual can be infinite. This is quite obvious because infinity is not an actual number but a continual process of adding one forever. And it's just as nonsensical as a flat earth to even imagine it might be negatively curved and have edges as others suggest.

The cosmic hypersphere is also clearly very much larger than the visible universe within the particle horizon because the 3-dimensional surface of the visible universe is flat within a small margin of error. However Universal Reality predicts it will be found to have a very slight positive curvature that confirms it's actually the surface of a hypersphere.

The evolution of the universe consists of the continual replacement of the current 3-dimensional outer spatial layer with a newly computed one in every P-time tick. In this way the radial P-time dimension is continually extended with every tick and this produces the Hubble expansion of the spatial surface.

However this geometry should not be taken too literally. Remember that spacetime is actually a logico-mathematical structure rather than a physical structure. The cosmic hypersphere is just the mathematical form the dimensional relationships in the entanglement network take at the cosmic scale.

COSMIC INFLATION

The expansion of the universe has apparently not been constant over time, as a constant extension of the P-time radius would suggest it should be. This is especially true of the inflationary period in the first fraction of a second after the big bang in which the size of the universe seems to have expanded exponentially from almost no volume at all to cosmic dimensions (Wikipedia, *Cosmic inflation*).

This initially appears to be a problem for the hypersphere geometry until we recall that spacetime isn't a physical universe but the computational result of particle interactions. Thus the universe isn't actually a geometric hypersphere whose radial extension *produces* a surface expansion. The hypersphere is simply an observational model mapping the dimensional data points at any particular P-time.

Spacetime is created by quantum events so as particles began to be created and interact in the inflationary period a spacetime volume sufficient to contain them would have been automatically created along with the particles. This in itself is sufficient to account for inflation.

Because clock time rates depend on the spatial velocity present the overall *cosmic clock time rate* in our barely expanding universe is currently close to c. However during inflation the enormous universal spatial velocity of expansion would have reduced the overall cosmic clock time rate to nearly zero. So in terms of our current look back clock time rate inflation seems to have occurred almost instantaneously even though it actually took very much longer. Inflationary expansion could have even taken place at sub-luminal rates but because clock time was flowing so much slower it seemed to take place almost instantaneously. Thus by the STc Principle inflation appears to have taken almost no clock time at all while innumerable P-time cycles were involved in computing it.

Thus whatever the source of inflation and the apparent current accelerating Hubble expansion it's quite clear that P-time and universal clock time tick rates need not be the same so a hypersphere remains the most reasonable shape of the universe. This is discussed in more detail in *Universal Reality, The New Theory of Everything* (Owen, 2016).

So in a computational universe the initial inflation of the hypersphere doesn't really require a physical explanation. The dimensionality necessary to incorporate all the newly created particles of the universe and their interactions is automatically created by their presence.

VISUALLY CONFIRMING THE HYPERSPHERE

It's also important to note that we can directly confirm the hyperspherical geometry of the universe with our own eyes. We actually see all four dimensions including the time dimension as distance in every direction from whatever point we occupy.

We obviously see the 3 orthogonal dimensions of space in all directions with our own eyes, and we also see the 4th P-time dimension as distance in every direction from any point. This is exactly what a 4-dimensional hypersphere should look like from the inside.

Of course we are actually seeing a slice called our past light cone through the 4-dimensional hypersphere but it's quite clear the inside of a hypersphere is what we actually see when we look into the night sky.

TIME TRAVEL & SPACE TRAVEL

Because the 3-dimensional spatial surface of the hypersphere is continually recomputed at every P-time tick, only the current surface of the hypersphere actually exists. The current present moment surface of the hypersphere is the entire actual universe and outside this surface there is nothing actual at all. All the past P-time surfaces have vanished forever and have no existence whatsoever except in their effect on the present, and no future surfaces have yet been computed and so cannot exist.

Thus only the current present moment exists, and the past and future don't exist. Therefore there is no actual past or future to time travel to or from. It's simply impossible to leave the current universal present moment because it's all that exists. The entire universe exists only in the current P-time tick that computes it. Thus the entire universe is the current present moment surface of the hypersphere and nothing else exists.

As to time travel we are all continually traveling through time at the speed of light on our own clocks. So in that sense we are all already time travelers. And we can of course travel through clock time at different rates in the present moment depending on how much spatial velocity we have.

So we can travel through time at different rates, which means we could potentially age much slower or faster than another observer. So given sufficient spatial velocity over time it would certainly be possible for a space traveler to depart ancient Rome and arrive here in the present not much older than when he started. But that is not actually traveling to the future; it's just traveling much slower in time in the present. Both the Roman space traveler and people back on earth were always in the same current P-time moments as P-time progressed from Roman times to the present. The Roman space traveler was just aging much slower because he was traveling much faster through space. Thus the space traveler's clock would have been ticking at a much slower rate than clocks back on earth during his trip.

Conversely it would be possible for us to embark on a space trip with very high spatial velocity and return to earth after centuries had passed. But this is simply because our high spatial velocity slowed our time and our aging during the trip. Thus theoretically we could plan our trip by adjusting its duration and spatial velocity to return to earth at any point in the 'future' at any (older) age we desired.

However this is not really a trip to the future because both we and the earth always remained in the same current present moment during the entirety of the trip. We were just aging slower than things on earth because we had a very high spatial velocity. And of course there is no going back to experience the intervening events on earth we missed because they have vanished into the past.

So there simply is no possibility whatsoever of any *time travel paradoxes* like going back to the past and killing your own ancestor. It's simply impossible to generate any such paradoxical contradictions since everyone remains in the same current present moment at all times. As much as we might like to it's simply impossible to change even the minutest detail of the past.

Because time passes much slower at relativistic velocities the STc Principle makes *space travel* much more feasible, at least theoretically, because it can greatly shorten the passage of time on the traveler's clock.

For example at only a 1g constant acceleration it would take only around 42 years to travel the 42,000 light years to the center of the galaxy on the traveler's clock (Misner, et al, 1973, p. 719). The huge spatial velocities generated by constant 1g acceleration would slow the velocity through time enormously as well and the trip could be accomplished within a human lifetime.

There are of course practical problems with such a space flight such as an effective propulsion source, and the problem of avoiding interstellar objects at near light speed, but at least it's theoretically possible.

As for time or space travel via worm holes I see no evidence they are even possible in a universe that exists everywhere in a common universal present moment. The uniform c velocity of all points in the fabric of spacetime also appears to preclude it.

LORENTZ CONTRACTION

Relativistic or Lorentz length contraction is a putative shortening of the length of an object predicted to occur in the direction of a relativistic motion. However there are a number of problems with the concept and it's better understood as a consequence of time dilation on spatial measurements instead.

Take the observed phenomenon of cosmic rays producing muons as they hit the upper atmosphere. Given the extremely short half-lives of muons they should all decay before reaching the earth's surface but they don't.

The standard explanation is that we observe this because the coordinate time of the muons is slowed due to their relativistic speeds and they have more time in which to decay and so are able to reach the earth. This is standard special relativity.

However some physicists claim that conversely in the frame of a muon time runs at the same rate as before thus to reach the surface of the earth before it decays the earth and its atmosphere must be Lorentz contracted in the direction of their relative motion so the distance traversed is short enough for the muon to reach the surface before it decays in its standard half-life (http://physics.ucr.edu/~wudka/Physics7/Notes_www/node79.html).

But the correct explanation is that it's the muon that has an actual velocity relative to the universal computational metric of the entanglement network and thus its proper time velocity is actually slowed while the earth's isn't. This is observably true from the perspective of an earthly observer's frame.

From the muon's perspective there are two opposing effects. Because the earth seems to be moving at the same speed relative to the muon as the muon is relative to the earth, the *coordinate time rate* of the earth appears to be slowed relative to that of the muon.

That is the *observational relativistic effect* from the perspective of the muon but the *actual relativistic effect* remains the same as before, simply that it's the clock time of the muon that's actually slowed because it's the muon that is actually moving at a high spatial velocity relative to the spacetime background with respect to which this is actually computed. Thus Lorentz contraction has nothing to do with it.

Basically the whole concept of Lorentz contraction is just an inferior way of looking at time dilation and textbook discussions confirm this by always explaining that calculating a length contraction depends on simultaneous measurements of the positions of the two ends, so the length always depends on measurements of times, which are of course different in different frames. The length contraction is then said to be due to the difference between simultaneous measurements of the ends in different reference frames.

This is true enough but actual versus observational time dilation is a much better way to understand this. So Lorentz contraction is just how measurements of length are affected by time dilation rather than being a fundamental aspect of relativity itself.

In fact it's been impossible to directly observe Lorentz contraction in any object long enough for its length to be measured due to the impossibility of moving large objects at fast enough relative speeds. It's only been measured in subatomic particles and it's always just an alternate interpretation of their time dilation.

In all cases Lorentz contraction is an *observational* rather than *actual* relativistic effect. This means that there never is any actual permanent contraction of any object that all observers agree upon once relative motion ceases as there is with elapsed clock time. Lorentz contraction is simply a way to understand how time dilation affects measurements of length.

In any case the best approach is to ignore Lorentz contraction and talk about time dilation instead. Then if we wish we can consider the effect of time dilation on *measurements* of length. Lorentz contraction is a subsidiary consequence of time dilation rather than a primary relativistic effect.

COSMOLOGY

A COSMOLOGICAL MODEL

So far we've arrived at a simple, elegant, and highly explanatory model of how the universe works. At the most fundamental level the universe and everything in it exists as data in a universal computational space where it's continually recomputed in every P-time tick. This computational space is non-dimensional or pre-dimensional in the same sense as a computer program defines a non-dimensional computational space where dimensional values are just numbers in computer memory.

Now computations of particle interactions produce dimensional relationships among those particles. And just as the data of a simulated universe on a computer exists as data in computer memory, but can be displayed as a 4-dimensional space (a 3-dimensional space evolving in time) in a virtual reality headset, so the actual numeric data of the universe in computational space can be displayed as a 4-dimensional world in the simulations of reality constructed by our brains.

Thus the universe actually consists of pre-dimensional data and it's the minds of observers that simulate and project it out around us as the dimensional space we appear to inhabit. How the data universe is simulated as a spacetime world is key to understanding the true nature of reality and how the universe really works.

Thus we simulate the numeric data of particle interactions in computational space as the 3-dimensional *surface* of a hyperspherical universe filled with its objects and beings. At every P-time tick all these entities evolve computationally in interaction with each other.

Both the particle structures of things and their dimensional relationships evolve together with every P-time tick. And it's their dimensional relationships that all together form the consistent logico-mathematical network our brains interpret as a physical spacetime in which they and we exist.

An important aspect of these dimensional computations is that they compute both the spatial and clock time relationships among things. Thus the individual clock times of different processes run at different rates within the same universal hypersphere surface based on the spatial velocities they have relative to the absolute computational metric in which they are computed. Finally if we zoom back from the data of all the individual interacting particles we see the human scale dramas they create emerge just like dramas on TV screens emerge from patterns of minute colored pixels. This enables a deep understanding of how the universe as a whole evolves computationally in a manner consistent with relativity.

As explained in the chapter on *Quantum Reality* we can easily incorporate quantum phenomena into this model simply by assuming each separate coherent process is being computed by a separate application of the universal processor each with its own random oscillation pattern in the calculation of space versus time velocities.

So each evolving coherent process can be visualized as a separate dimensional layer of the current surface of the hypersphere being computed by a separate application of the universal processor. This is what we observe at the quantum scale as wavefunctions dimensionally indeterminate with respect to each other including those of our observations.

However at the classical scale these separate layers of the current surface are too small to be resolved and we see the hypersphere as a single evolving surface. In this way our computational model incorporates both relativity and quantum theory in a unified manner as the human scale observable universe emerges from individual particle interactions.

A NEW DARK MATTER THEORY

The existence of an invisible form of matter called *dark matter* was first proposed to explain observational anomalies in the motion of stars rotating in spiral galaxies. Observations suggest that galaxies rotate as if they had halos of invisible mass around them because their outer arms are rotating faster than would be expected based on their apparent masses. The amount of dark matter necessary to explain the movements of galaxies is huge, about 5 times the amount of visible matter in the universe (Wikipedia, *Dark matter*).

Dark matter has been sought in the form of various types of new particles but so far none have been found. However Universal Reality suggests another possible explanation for the dark matter effect, which seems to be original to the author's 2013 book *Reality* (Owen, 2013). This proposal is a simple and rather obvious consequence of the progressive Hubble expansion of the universe.

The Hubble expansion is an expansion of the relatively empty space *between* galaxies and galaxy clusters which makes up most of the universe. By contrast the space *within* galaxies isn't expanding because it's gravitationally bound by their mass (Misner, et al, 1973, p. 719). Thus the earth, the solar system, our galaxy, and we are not expanding but the space between galaxies is expanding. This uneven expansion is obviously true because if everything was expanding uniformly the expansion wouldn't be observable.

The result is an uneven Hubble expansion that warps space around the boundaries of galaxies; precisely in the area that dark matter is expected to be found! And from general relativity we know that any warping of space will manifest as a gravitational field. Thus we have a natural explanatory mechanism for the dark matter effect that involves only the expected warping of space from the uneven Hubble expansion around galaxies and doesn't require the existence of any new particles.

This warping may or may not be the cause of the entire dark matter effect, but it certainly should be producing a very large gravitational effect, since the uneven expansion over the lifetime of the universe should produce very large spatial warps around galaxies.

Distributions of dark matter can be mapped by tracing gravitational deviations in the expected paths of light beams from sources beyond them as well as its gravitational effect on local visible matter. These maps indicate a distribution of dark matter generally around galaxies but sometimes offset as well. However there is nothing to prevent these Hubble space warps, once they are created, to have a life and movement of their own. Thus dark matter distributions should initially form as halos around galaxies and galaxy clusters but then be able to move as massive objects on their own due to their gravitational interactions.

Once Hubble warps are formed they are effectively *free gravitational fields* that can move through space just as galactic masses do. The continued existence of a dark matter mass is not dependent on the original galaxy it was created from. There will be a continuous creation of new dark matter warps around galaxies, but once created these can trail away and should leave detectible plumes of warping behind that reveal how galaxies have moved over time.

Over the course of the expansion of the universe the actual effects will be extremely complex because the distribution of galactic matter with time is extremely complex. It should be fairly easy to test at least the viability of this theory by comparing the current distributions of dark and visible matter and inferring their relative motions over time and making a calculation of whether the expected warping would account for the gravitational effects of known dark matter concentrations.

This is one possible explanation of the dark matter effect, but not necessarily the only one. Nevertheless there should be a very substantial warping due to the uneven Hubble expansion, and that warping should be producing quite a large gravitational effect. Where is that effect if it isn't the dark matter effect? It must show up somewhere. The evidence seems quite strong and it certainly simplifies things by not requiring any new unknown types of particles beyond the Standard Model.

This theory of dark matter also neatly explains why dark matter is dark. Not being an actual form of particulate matter it obviously doesn't emit light. Thus it's naturally invisible and interacts with regular matter only via the gravitational force, as dark matter is known to do.

The theory also implies that the amount of dark matter should increase over time with the expansion of the universe and the continuing uneven expansion of space around galactic masses. There is in fact recent evidence this may be true lending further credibility to the theory (http://www.eso.org/public/usa/news/eso1709/).

Not only that but the increase in dark matter should be gradually exponential as the increasing presence of more and more dark matter adds to the total positive gravitation of the universe resulting in ever more warping around it which in turn manifests as increased positive gravitation in a continuing feedback cycle.

Thus depending on the balance of dark matter and dark energy there could be an automatic mechanism that brakes the Hubble expansion and ultimately produces enough dark matter to begin to reverse expansion and initiate a gradual collapse into a universal black hole.

BLACK HOLES

By the STc Principle the maximum space plus time velocity of any point in space is c. This means that a black hole will be a solid sphere of maximally packed mass where the velocity of time is zero and the intrinsic spatial vibrational velocity of gravitation is c. This corresponds to the maximum possible gravitational field, the field of maximum possible intrinsic vibrational velocity.

Thus Universal Reality avoids any paradoxical singularities at the center of black holes where the laws of physics break down, a problem that continues to plague standard interpretations of relativity. As more and more mass is added to a black hole it simply increases its volume forming a larger and larger solid sphere where all the fixed constant c velocity of spacetime takes the form of vibrational spatial velocity and the velocity of clock time has slowed to a stop.

Thus, contrary to physics orthodoxy, the interior volume of black holes can't collapse to a singularity and instead increases with the addition of mass. The notion of black hole singularities is based on the traditional interpretation that space is infinitely compressible, which is impossible given the STc Principle. Thus black holes should be solid rather than the largely empty structures envisioned by current misinterpretations of relativity (Thorne, 1994).

So inside a black hole clock time is essentially frozen and nothing appears to happen, even light is frozen in place and can't escape. The speed of light relative to the computational background is zero as all spatial velocity is that of the intrinsic vibrations of gravitation even though locally it's still c as always as clock time has also stopped.

Moving outside the surface time begins to flow and light begins to travel through space again faster and faster with increasing distance from the event horizon. In all cases light is always traveling at the *local* speed of light, which is actually just the local speed of clock time. Both are just equally much slower near the black hole than in empty space.

In addition to the effect on the fabric of spacetime we must also consider what happens to particles in black holes. As gravitation increases sufficient vibrational energy is available to overcome the mass deficit that normally keeps electrons and protons from combining into neutrons and atoms first collapse to the size of their nuclei as neutron stars are formed.

This normally occurs when stars of sufficient mass burn out and gravitation overcomes outward radiation pressure producing a collapse. But it will also occur as a result of sufficient mass accretion as black holes are formed and grow.

Mass can never quite attain the speed of light because mass itself is a form of vibrational spatial velocity and the total spatial velocity can never exceed c. So there is a question of what happens to mass as the intrinsic spatial velocity of the gravitational field approaches c in a black hole. Perhaps it's converted to electromagnetic waves, which do travel at c but stand still as frozen light inside black holes? The amount of gravitation wouldn't be affected if the mass content were converted to an equivalent amount of pure photon energy.

What happens to particles inside black holes is probably best modeled by running the big bang backwards. In this view even neutrons will eventually be reduced to their constituent quarks and gluons. Whether beyond that there is some point where even particle components dissociate and are compressed back into the fabric of spacetime from whence they originally emerged is an open question. The maximal positive pressure of the black hole on space might well have this effect.

As Universal Reality models all types of charges as various forms of spatial velocity the maximal gravitation field of black holes might compress all the particle components back into their particular fields as various virtual spacetime components thus completing the transformation of the spacetime fabric into areas of zero clock time c valued spatial velocity as all particles are absorbed back into the fabric from whence they originally emerged in the big bang.

If this is correct then technically the mass inside a black hole would vanish but the gravitational field certainly remains. What remains of the black hole could just be another type of free gravitational field similar to that proposed in the preceding section to explain dark matter.

One might think that if time stops there can be no vibrational spatial velocity. But this is true only in relativistic frames. Just as linear velocity relative to the fabric of spacetime can increase indefinitely towards c even as its local clock time slows to a crawl, so the vibrational velocity of a gravitational field is with respect to the absolute computational background where P-time proceeds at the same intrinsic rate as always. Clock time slows to nothing but P-time proceeds at the same universal rate computing it all including the slowing of clock time.

As previously discussed all processes in the universe are subject to quantum scale random oscillations between space and time as they are computed and this must also be true of the time and space within black holes. Because of these random oscillations there is never a complete conversion of time to spatial velocity. There will always be some quantum scale temporal velocity within black holes at least at the scale of the zero-point energy.

If this is true then the average intrinsic spatial velocity of black holes could be ever so slightly less than c. This might allow a slight amount of radiation to escape from black hole surfaces and their rate of evaporation could be significantly greater than by Hawking radiation alone though still extremely slow (Wikipedia, *Hawking radiation*). This could conceivably be a testable prediction of our theory.

It's important to understand what happens to an object as it falls into a black hole. One might expect it to fall faster and faster reaching nearly the speed of light as it approaches the event horizon. As time slows in the object's frame this is true, however what is really happening is that the space the object traverses becomes more and more densely packed with the intrinsic vibrations of the black hole's gravitational field. Thus the object has to traverse their increasing peaks and valleys as it falls. (In the usual interpretation spacetime becomes increasingly curved).

The result is that the object falls faster and faster but the actual traversal distance through space increases even faster. As a result an object falling toward a black hole actually appears to slow down as it approaches the event horizon. The light returning back to an observer out in space also has to traverse the peaks and valleys in reverse and this increases the effect. The light from the object also increases in wavelength towards the red and beyond and eventually stops due to the maximal gravitational red shift and the inability of light to escape the event horizon.

Some physicists suggest infalling objects appear to pile up on the event horizon but this is incorrect because as objects approach the event horizon their light both becomes fainter as the number of photons per second escaping the black hole's gravity approaches zero, and their frequencies redden into invisibility due to gravitational red shift. For both these reasons their images fade into invisibility as they reach the event horizon.

Thus there is no pileup of objects on the event horizon and the derivative notion of the event horizon somehow being a 2-dimensional hologram of the information within the black hole are both nonsense. Likewise I see no merit to the entire related holographic hypothesis that all the information of the universe somehow exists as a 2-dimensional hologram around some nonexistent 'edge' of the universe and that the information internal to the universe is somehow a projection of this 2-dimensional holographic surface (Wikipedia, *Holographic principle*).

An observer falling with the object itself notices no slowing because his clock time is slowing to a crawl and he seems to be falling faster and fester But since time stops at the event horizon his experience also stops and nothing more happens.

Due to the extreme gravitational gradient particulate objects are ripped apart into their component particles before they reach the event horizon and likely into a particle component plasma within it. So black holes are likely composed of particle component plasma.

The slowing of linear velocity in strong gravitation fields also ensures that an object's total spatial velocity can never exceed c. If linear spatial velocity and gravitational spatial velocity were additive that could happen which would violate the STc Principle. An object traveling at near the speed of light entering the vibrational spatial velocity of an intense gravitational field could have a total spatial velocity greater than c if that was true.

Instead the linear spatial velocity must traverse the greater and greater spatial vibrations of the gravitational field, which slows its linear velocity compared to what it would have in empty space. Thus the total spatial velocity of the object can never exceed the speed of light and the STc Principle is preserved. This is equivalent to the standard relativistic model of objects taking longer to traverse gravitationally curved space and near forever in the intensely curved space around black holes.

This analysis also highlights the difference between actual and observational relativistic effects. Actual relativistic effects happen locally where they are computed while observational relativistic effects are those viewed non-locally from within relativistic situations where spatial velocities and thus time velocities and relativistic circumstances are different.

Of course actual relativistic effects are being computed at all locations in the universe but the actual event being computed out in space is an *observation* of the object falling into the black hole, and the observation is not what's actually happening to the infalling object. The actual computation of the object falling into the black hole is in fact the *object actually falling* into the black hole.

They both appear equally real to local observers and are correctly described by the equations of relativity but the true picture is only that of the metric in which events are actually being computed rather than just being observed. Individual observers are fooled by tricks of light moving at finite velocity through areas of different spatial velocity. Things are always computed where they happen, and the computational metric in which events are computed is the privileged background metric in which they actually occur.

This new theory of uniform solid black holes without singularities could have testable consequences. If so that could be an important test of the theory of Universal Reality.

DARK ENERGY

Dark energy is a phenomenon postulated to account for the apparent gradual acceleration in the Hubble expansion of space over the second half of the age of the universe. The density of dark energy ($\sim 7 \times 10^{-30}$ g/cm^3) is extremely low, much less than the density of ordinary matter or dark matter within galaxies (Wikipedia, *Dark energy*). However it comes to dominate the mass–energy of the universe due to its apparent uniformity across all space.

Theories of dark energy fall into two main categories; the *cosmological constant*, which is constant across all space, and *quintessence theories*, which postulate non-universal variations in the rate of the Hubble expansion, usually particle based. Of these the cosmological constant seems more likely because it's already contained in the Einstein field equation and doesn't require the existence of any new undiscovered particles beyond the Standard Model.

Positive pressure is one of the components of the stress-energy tensor that accounts for gravitation in the Einstein field equation as explained in *Relativity Made Easy* (Owen, 2016, Note 3). For example there is positive pressure on space in intense gravitational fields that adds a very small gravitational effect.

So what's needed to account for dark energy is a *negative pressure*, which would produce a repulsive gravitational effect that would tend to expand space. An example of negative pressure is radiation pressure emanating outward from a source, which tends to expand space as opposed to a positive gravitational pressure that tends to compress it. Locally both these pressures are negligible and can have appreciable gravitational effects only if they extend over vast interstellar distances as the cosmological constant is theorized to.

The cosmological constant is a property of the fabric of space itself. It could be a negative pressure similar to radiation pressure without the radiation. On the other hand the expansion of space could be associated with an emission of radiation from the creation of new particles from the fabric of space.

This association of the expansion of space with the release of radiation and a positive cosmological constant is reminiscent of the *Unruh effect* in which acceleration relative to the spacetime background produces an observational release of particles of radiation out of empty space (Wikipedia, *Unruh effect*). So the expansive velocity of space could result in a production of particles of radiation from the fabric of spacetime that would produce the necessary negative pressure.

It's generally accepted that the cosmological constant is an effect of the zero-point energy of the quantum vacuum. And in our theory the zero-point energy is due to random processor cycle oscillations in the processor that continually computes the universe. In this model all that would be needed to explain the accelerated expansion is a minute variation in these oscillations.

If this is correct then dark energy could be as simple as a very slight increase over the second half of the universe in the processor cycle oscillations. The dark energy expansion is extremely minute and the processor cycles would have to increase only infinitesimally to produce the apparent Hubble acceleration.

This might theoretically be testable as it should produce extremely minute variations in the form of wavefunctions over time but those would seem to be far below the resolution of experiment.

There is also a well-recognized problem in trying to correlate the zero-point energy with the cosmological constant. The value of zero-point energy turns out to be 120 orders of magnitude greater than it should be to account for the cosmological constant, a problem known as the vacuum catastrophe (Wikipedia, *Vacuum catastrophe*).

Both the zero-point energy and accelerating expansion values appear to be soundly based in observational evidence, so it's reasonable to conclude they can't be one and the same thing. The logical conclusion is that only one aspect of the zero-point energy accounts for the cosmological constant. This could simply be a minute discrepancy in the rates of virtual particles appearing out of the quantum vacuum over those disappearing back into it in the zero-point energy. These additional newly actualized particles would provide the needed negative radiation pressure necessary to account for the Hubble expansion, and could well vary over time in accord with the Hubble expansion and a minutely increasing cosmological constant.

The ultimate source or at least the correlated source of dark energy would be the rate of continuing extension of the radial P-time dimension of the cosmological hypersphere. One can think of this as stretching the fabric of spacetime thereby reducing its density causing new particles to pop into existence whose increased radiation pressure would be consistent with the expansion and the cosmological constant. In this manner a constant zero-point energy density of space is maintained as it expands, and the addition of new particles of radiation to expanded space would add to the total negative pressure resulting in the gradual acceleration observed. It all would fit together neatly as different aspects of a single process.

In this way the zero-point energy density of the fabric of spacetime would remain nearly constant as space expands, and the zero-point energy and cosmological constant could have vastly different values but ultimately be aspects of the same phenomenon due to the manner in which the universal processor computes the extension of the P-time radius of the cosmic hypersphere.

Another consideration is that it's also not clear how accurately Hubble expansion rates are known over time. They are based primarily on observations of type 1a super nova standard candles, which exist only for short periods of time and few and far between, and there are none close enough to accurately determine the actual current expansion rate of the universe. In addition the current state of the universe is only locally observable so its impossible to have any idea of what is currently happening at any significant distance. See *Universal Reality* for a discussion of some of the problems involved (Owen, 2016).

Another possibility is that dark energy is just an effect of the increasing dark matter content of the universe. This is explained in the next section.

A POSSIBLE BIG BOUNCE UNIVERSE

We now arrive at a *possible* mechanism that would automatically generate an eternal cycle of big bounces in which the universe undergoes initial inflationary big bangs, expands over billions of years gradually increasing its dark matter gravitational mass till it eventually overcomes the expansion and begins a slow gravitational collapse to a universal black hole that instantly explodes into a new white hole inflationary big bang. This part of the theory is speculative but appears reasonable. The mathematical details need to be examined to see if they make sense.

The process works like this:

1. As explained in the section on dark matter the ongoing expansion of the universe continually produces new warps in the fabric of space around galaxies due to the uneven expansion between galaxies and intergalactic space.
2. By relativity any distortion of space acts as a free gravitational field so these warps are responsible for the *dark matter effect* around galaxies.
3. These dark matter warps add to the total size and gravitational mass of galaxies so even more warping is produced as space continues to expand. This process produces a slow exponential increase in the total gravitation of the universe that in turn slowly accelerates the production of additional dark matter.
4. The ever-increasing amount of dark matter progressively slows the expansion that produces it. Thus the expansion of the universe eventually comes to a halt and reverses.
5. However there is also an apparent opposite *dark energy effect* that currently seems to be accelerating the expansion. So the ultimate fate of the universe appears to depend on the balance of these two effects.
6. However it's possible the supposed dark energy acceleration of expansion may be an illusion in full or at least in part. To illustrate why assume a set of standard candles stretching away from earth at fixed intervals. Each candle has the same intrinsic brightness but their apparent brightness diminishes with distance.

Each candle also has a red shift that indicates how fast space at that distance is receding from earth. The red shifts of all candles together gives us a picture of how fast space was expanding at different distances and times over the history of the universe.

7. However the actual distance light has to travel from candle to candle depends not just on the nominal distance but on how strongly space is warped in between them. Thus if the dark matter warping of space is increasing over time light will take longer to traverse the nearer more recent intervals than the farther earlier ones. And because the stretching of space also increases the red shift of light that traverses it the apparent red shifts of nearer equally spaced candles will be greater than expected. And this will be mistakenly interpreted as a current acceleration of expansion that may not actually be happening.

8. This means that since *dark matter* is unevenly distributed rather than a universal phenomenon the apparent *dark energy* acceleration could be different at other locations and considerably less or perhaps even nonexistent if viewed from intergalactic space with minimal dark matter content.

9. Thus it's possible that the apparent dark energy acceleration of expansion may not actually be happening. It may be in whole or at least in part an artifact of the ongoing increase in dark matter lengthening the path light takes across space (increasing spatial curvature) in more recent times. And since dark matter is increasing over time that would also be consistent with the apparent increasing Hubble acceleration. To what extent this may be true remains to be determined.

10. So assume for the moment that the current expansion of the universe is simply due to the initial momentum of the big bang and inflation and if there is a dark energy expansion that the increase in dark matter eventually overcomes it. Eventually the increasing dark matter gravitational mass of the universe as it expands overcomes its expansion and initiates a universal contraction.

11. During the period of contraction clock time continues to run forward in the same direction as before as it continues to be computed by the constant progression of P-time that computes the entire cyclical process. However clock time eventually begins to slow as the universal gravitational field strengthens as the contraction progresses and more and more of the constant c valued space plus time velocity of the universe is expressed as gravitational spatial velocity rather than clock time velocity.

12. As the contraction progresses the night sky and universe as a whole gradually turns from dark to light. Since no part of the

universe is now expanding away from any other faster than the speed of light particle horizons disappear and the entire universe becomes visible. The light from all galaxies in currently hidden areas of the universe gradually becomes visible and fills the night sky with light. The night sky becomes as Olber imagined it, not because the universe is infinite but because it's the continuous surface of a contracting hypersphere (Wikipedia, *Olber's paradox*).

13. As the universe contracts further galaxies gradually move closer and closer and begin to collide. As galactic mass condenses a new era of star formation begins in which generations of increasingly large and short-lived stars are formed, become supernovae, and increasingly become large enough to form a new generation of stellar black holes.

14. Increasingly galaxies become populated with black holes instead of stars that in turn collapse into their central supermassive black holes leaving a universe of supermassive black holes instead of galaxies. These black holes become increasingly numerous as the universe continues to contract and then begin to combine into larger and larger hyper massive black holes which eventually all collapse into a single universal black hole. The blinding light that had grown to fill the universe subsides into universal darkness again.

15. Due to the STc Principle this universal black hole is not a singularity within an event horizon but a finite uniform sphere within an event horizon containing the entire mass-energy of the universe. And because there are no more particles outside it to maintain a supporting external dimensional spacetime, spacetime itself also collapses into the black hole. The entire observable universe has now become a maximally compacted single universal black hole.

16. Like all black holes this is a sphere of c valued gravitational spatial velocity within which clock time stops. The entire universe has become a relatively small uniform sphere of the maximum possible field of attractive gravitation within which nothing happens.

17. Now to produce a new black hole to white hole big bang transition all that's needed is a mechanism to convert its gravitational *vibrational spatial velocity* into the *linear spatial velocity of heat.*

18. This is consistent with the accepted model of the big bang, which is heat driven. In the accepted model the big bang begins with enormous heat energy and as heat decreases with the resultant inflationary expansion the universe undergoes a sequence of

phase transitions as the four forces separate and more and more complex particles reach energy levels low enough to form and become stable (Wikipedia, *Big Bang*). From this point the phase transitions of the big bang more or less follow those outlined in the standard theory with a few additions.

19. The necessary mechanism has to do with what happens to massive particles inside the universal black hole. The maximal gravitation of the black hole stops clock time, particles cease to interact and since particle interactions create dimensionality spacetime collapses within the universal black hole.

20. All compound particles also dissociate since they are dimensional constructs. In particular protons and neutrons dissolve into quark-gluon plasma.

21. This immediately converts 99% of the mass into free vibrational velocity since the total mass of quarks and gluons is only around 1% of the mass of the nucleons they compose (Wikipedia, *Quark#Mass*). The rest of the mass of nucleons comes from the vibrational energy of their quarks and gluons bound by the strong force. Thus when nucleons dissociate into quark-gluon plasma at least 99% of their vibrational spatial velocity is instantly converted into the linear velocity of heat or a close equivalent.

22. As all mass and particles disappear this unbinds the gravitational spatial velocity of the black hole instantly converting it to an equivalent amount of the linear spatial velocity of heat that immediately produces a new white hole big bang that inflates the universe again beginning a new big bounce cycle.

23. Simply stated the total spatial velocity of the universal black hole instantly transforms from the vibrational spatial velocity of gravitation to the spatial velocity of heat. The entire gravitational mass of the black hole is instantly converted by $e = mc^2$ to near c average velocity heat in a hyper massive nuclear explosion. The negative pressure of this heat acts as the repulsive force that instantly inflates the new universe creating new particles out of the quantum vacuum fabric of spacetime in a runaway inflation. As a result the universe reinflates in an enormous cosmic explosion as the spatial velocity of the entire gravitational mass of the universe is instantly converted into heat energy.

24. This heat energy initiates a new big bang and provides all the momentum that drives inflation and the subsequent Hubble expansion of the resulting universe.

25. So essentially the universal black hole to white hole conversion is a universal nuclear explosion in which the entire mass of the universe is instantly converted into heat energy. Ordinary nuclear explosions are particle interactions in which very small

percentages of very small amounts of mass are converted into heat and radiation. They occur when particles dissociate into their particle components to form new particles in which some of the mass can't be packed into the new particles. The only way the energy of the excess mass can be conserved is if it's converted to other forms of spatial velocity. This converts some of the mass into heat and radiation energy creating a nuclear explosion.

26. So to produce the necessary nuclear explosion of a universal black hole to white hole transition all particles dissociate into their particle components and all their mass is converted to heat energy. This happens because all particle interactions stop due to the cessation of clock time. So if clock time stops particles dissociate into their particle components. This may happen within all black holes. But in a universal black hole the dimensional spacetime of the entire universe collapses as well and this erases the dimensionality of the black hole's gravitational field converting it to heat and radiation energy.

27. Local black holes are maintained as dimensional objects in space by the particle interactions of the rest of the universe around them. But if all the particles in the universe fall into a universal black hole the spacetime outside is no longer maintained by particle interactions and vanishes. Only in this case does the residual gravitational field of the black hole dissolve as well with its spatial velocity converted into heat in a white hole big bang explosion. This in turn provides the energy to actualize particle components out of the quantum vacuum into new particles by jostling them together at sufficient linear velocities. This begins to actualize the particles of the next cycle in a new big bang and inflation.

28. Ordinary black holes maintain their gravitation and don't explode into white holes so their maximal c valued gravitational vibrations in themselves isn't sufficient to produce white holes, otherwise we'd see white holes exploding out of local black holes all across the universe. So it must be the collapse of the dimensional universe itself into a universal black hole that converts it into a white hole big bang. And it can only be the continuing particle interactions outside local black holes that maintains their dimensionality and prevents their transition to white holes. Local black holes don't become white holes because their spatial fabric is maintained by the ongoing particle interactions of the rest of the universe. Only if dimensional space itself collapses into a universal black hole can it transition to a white hole.

29. Since it's particle interactions that dimensionalize the quantum vacuum into what we call spacetime, as new particles are created

by the new big bang and begin interacting their interactions automatically create a new dimensional spacetime universe large enough to contain them.

30. This in itself is sufficient to explain inflation and is consistent with the subsequent expansion of the universe as simply the initial momentum produced by the heat of the big bang. As the universe of particle interactions expands enough dimensional spacetime to contain them is automatically created and this is the source of the Hubble expansion.

31. During inflation the spatial velocity of the entire universe begins near c. This means the universal clock time rate during inflation was near zero. Thus cosmic inflation appears to have been almost instantaneous from our look back perspective. However in the frame of the expansion itself inflation would have taken enormously longer. Thus cosmic inflation could well have taken very much longer than we think it did.

32. If this is true then cosmic inflation wouldn't need its hypothetical *inflaton field* (Wikipedia, *Cosmic inflation*). And instead there could have been a smooth heat driven diminishing expansion from the very beginning of the big bang to the present. The expansion and drop in universal spatial velocity would progressively increase the velocity of universal clock time resulting in a smooth gradually slowing expansion as P-time continually ticked at the same rate. This would greatly simplify the cosmology of the early universe and make it much more reasonable.

33. The conversion of maximal spatial velocity from that of gravitation to heat also automatically reverses the entropy state of the universal black hole from maximal to minimal as explained in the section on *The Arrow of Time* above. The *maximum* entropy state of a universal black hole is all particles packed together as closely as possible. But in a gravity free heat expanding white hole this instantly becomes the *minimum* entropy state because the maximum entropy state is all particles as far from each other as possible. This solves the supposed problem of an initial improbable state of minimum entropy at the big bang (Penrose, 2005).

34. Through this entire process clock time doesn't reverse. It continues to be computed as before by the continuing forward progression of P-time even as the universe collapses and reinflates. However clock time effectively slows to a stop in the universal black hole and then restarts and slowly speeds up in the new big bang universe as the universal spatial velocity of heat decreases. Thus universal clock time slows to a stop in the big bounce and gradually speeds up to its eventual maximum rate as

the universal expansion slows to a contraction. And conversely total spatial velocity reaches a maximum on both sides of the big bounce and slows to a minimum as the universe transitions from expansion to contraction. In this way the total c valued space plus time velocity of the universe always remains equal to c in accordance with the STc Principle. So the whole cyclical big bounce process can be seen as a continual conversion of the total constant c valued energy of the universe from spatial velocity to clock time velocity and back again.

35. Thus we have natural mechanisms that automatically generate the two turning points or antipodes of a big bounce cycle that are direct consequences of the expansion and contraction and simple fundamental principles of Universal Reality. The expansion itself automatically produces the increasing gravitational attraction that eventually brings it to a halt and reverses it. And the formation of a universal black hole containing all particles and spacetime automatically converts the gravitational energy that produced it into the heat energy that destroys it and initiates the inflationary big bang of the next observable universe.

36. From this point the expansion creates a new expanding universe that produces a stochastic variation of the overall evolution of the previous observable universes in accordance with the complete fine-tuning. A process of *convergent emergence* begins wherein the observable universe evolves a new variation on the underlying plan baked into the complete fine-tuning.

37. In each big bounce cycle the general principles of emergence remain the same but produce different statistical results due to the quantum randomness intrinsic to all particle computations.

38. An interesting question is whether the complete fine-tuning could be somehow refined and retuned by the evolved results of the previous cycle as its information is condensed in its collapse. If so the universe could perhaps be in an unending process of *continual retuning* through each of its incarnations towards some ultimate state of maximum efficiency, elegance, and even perhaps of self-awareness or self-realization.

39. This could conceivably happen if the highest hierarchy of emergence emerged some purposeful direction to its evolution and this somehow informed the complete fine-tuning during the next black hole-white hole transition to subtly reprogram the complete fine-tuning to produce this more directly, efficiently, and completely in the next cycle. Perhaps possible if the total information of the emergent evolution of each cycle is somehow condensed by its collapse into the complete fine-tuning of the next cycle.

40. The possibility is that the entire program of the universe could eventually evolve purpose just as the programs of individual biological organisms have. It then might gain the ability to exercise some control over both the virtual data of the complete fine-tuning as well as it's own program. At the highest-level of the hierarchy of emergence if the universe becomes purposeful and intelligent it might be able to even reprogram the seeds of its next incarnation. Each incarnation might bring it closer to an ultimate universal self-awareness, consciousness and realization.

41. By analogy this could be considered a sort of epigenetic mechanism tweaking the expression of the DNA of the universe's complete fine-tuning. In effect each incarnation would select the most well adapted emergent information structures and programs and this information is then folded back into the complete fine-tuning of the next incarnation.

42. There is clearly much still to be discovered and the observable universe may continue on forever in an unending evolution towards some ultimate perfection!

If this theory is correct we could be living in the expansion phase of a continuous adiabatic big bounce universe that eternally cycles between universal expansion and contraction punctuated by black hole to white hole transitions. In an adiabatic universe nothing is lost to the exterior because there is no exterior to the universe.

There are a couple of complications that have been ignored here. First the universe is subject to quantum oscillations as it's computed so the c spatial velocity of black holes is never quite complete. There will always be a small random component of clock time velocity. How this would affect a putative big bounce cycle needs to be further examined.

Also the collapse is not direct but will be slowed by the angular momentum of massive objects around black holes as they form. This leads to periods of orbital frictional radiation of energy back out into space but an eventual collapse is inevitable.

In conventional cosmology there isn't enough time in the early universe for the supermassive black holes at the center of galaxies to form. In our model there will be plenty of time due to the very slow rate of clock time in the early universe. This is evidence in support of the theory.

While this big bounce theory is speculative, no matter what happens to the observable universe existence itself should continue to exist eternally as explained in the chapter on *Existence & Consciousness*. It's difficult to believe this could be true without it continually manifesting some sort of observable universe.

INFORMATION COSMOLOGY

Now we have a very clean model of our evolving universe that generates important additional insights.

The actual universe consists entirely of the current outer layer of the hypersphere. This outer layer is always directly computed from the data of the previous layer.

These computations all occur at the quantum level and thus involve random choices among probability distributions. In this way the future (upcoming) data state of the universe is always probabilistic rather than deterministic.

However once these random choices are made they cannot be changed even in the slightest detail. Thus the past is completely fixed and deterministic from the perspective of the present, which is the only actual perspective. This complete exactitude applies not just to the immediately preceding data state from which the current data state was computed but to every previous data state all the way back to the big bang and complete fine-tuning.

Thus the entire past could only have been exactly as it was to result in the present being exactly as it is. There is no way the past could have possibly been different in the slightest detail than it actually was.

Thus the complete past - present information state of the universe forms a completely deterministic fixed and unalterable theoretical structure totally consistent in both temporal directions. And nothing about this entire structure, not even the minutest detail, could have possibly been different than it was. Not even a single one of the uncountable myriads of random quantum events over the entire history of the universe could have possibly been different than it was.

The actual exact current information state of the universe

conclusively and absolutely falsifies any other possible past than what resulted in the present. There is only one single possible past that results in the present in every exact detail of the entire universe. There are no other possible alternative pasts in even the slightest detail. The entire past is completely and deterministically fixed in every minute detail.

In common parlance we often speak of the past as if it might have been different, and this is a useful tool in learning to predict future events. We imagine different possibilities in the past and consider the differences they could have made in the present to better understand the workings of reality and the effects of our choices in the future. But what does it actually mean to say something existed (past tense) in the past? All this can ever mean is that this present concept is part of an internally consistent mental model of reality in the *present*.

Being able to imagine different possible past events doesn't mean the actual past could have been any different than it actually was. It couldn't have been different because the actual past is quite clearly impossible to change even though our knowledge of it improves. What we really mean when we speak of alternative possible past events is that we can devise a *similar* event in the *present*. But of course we can never actually change any past event. It's completely impossible to exactly repeat any past event because all information states are connected with all the information of the entire universe as it was at the time the original event occurred and by definition those are now different.

Thus the entire past back to the big bang is completely and exactly determined by the current information state of the universe in every last detail and simply could not have been any different whatsoever and that includes the actual results of every random quantum computation.

The entire notion of different possible past states is an illusion based on our ability to repeat conditions *similar* to past conditions in the present on a *local* scale. Thinking in this way is convenient but applies only to experiments that are actually possible. It's certainly not possible to go back to the past and actually change anything and it's certainly not possible to recreate a different origin of the universe. Thus alternative possibilities have no meaningful application to the actual cosmological past. Alternative possibilities in the past are meaningless nonsense. This enables a new understanding of the anthropic principle.

BEYOND THE ANTHROPIC PRINCIPLE

There are several versions of what is called the Anthropic Principle but basically it's the contention that the fact that our universe is designed so that intelligent life has evolved is to be expected since only if it were true could we ask the question of why it was true (Wikipedia, *Anthropic principle*). Thus the fact that it's true in our universe is to be expected

However since most cosmologists think a universe that evolves intelligent life is statistically extremely improbable they then make the enormous leap that that there can be expected to be myriads of other universes in which it isn't true. From this all sorts of multiverse and bubble universe theories arise for which there is no evidence at all (Vilenkin, 2006) (Susskind, 2006).

The reason cosmologists think a universe that evolves intelligent life like ours is improbable is that the virtual data of the complete fine-tuning must be very precisely tuned for it to be true. And since they can think of no reason for the fine-tuning to be as it is they assume it's a statistical accident and that all other possible fine-tunings must be equally probable, and incredibly that each must therefore have existed and actually resulted in its own universe!

But this is a completely unwarranted assumption. First there are a number of reasons why any complete fine-tuning must be at least partially as ours is that are covered in my *Universal Reality, The New Theory of Everything* (Owen, 2016).

But second the fact that the existence of the universe as it actually is in the current present moment conclusively falsifies any possible difference in the complete fine-tuning in our universe, and the fact that there is absolutely no evidence at all of any other universes strongly suggests that the complete fine-tuning of our universe is the only complete fine-tuning possible.

And since the presumed possibility of alternative fine-tunings is the rationale for most of the currently popular multi-verse and bubble universe theories these are all immediately thrown into doubt (Wikipedia, *Multiverse*).

When one is presented with an unexplained fact such as the seemingly irreducible complete fine-tuning the logical approach is simply to try to find the reason for it or to accept it as fundamental. Certainly to imagine it implies the existence of myriads of entire other universes is the height of irrational implausibility!

Thus we can reasonably assume no other complete fine-tunings were even possible absent any evidence for them. Certainly there couldn't possibly have been any in our universe otherwise the universe would be different than it actually is which is impossible. So from the perspective of the present there is no reason to believe any other universes exist or ever existed, or that the complete fine-tuning could have possibly been different that it was and is in even its most minute detail.

Thus the past is completely fixed in every minute detail. In this sense the present determines the past. The exact present determines a past that would computationally produce the exact present as it actually exists. Thus the entire present-past is a completely deterministic logico-mathematical structure completely consistent in both temporal directions that cannot be altered in any detail whatsoever.

So it's only the future that is probabilistic since it's being actively computed by quantum processes. Different possibilities are continually being chosen at the particle level and these aggregate into computational choices being made among things at the classical level as well. Thus the future is probabilistic but once it's computed the entire past through present is immutably fixed and there is no possibility whatsoever that it could have been any different that it is and was all the way back to the complete fine-tuning.

FROM CAUSALITY TO CONSISTENCY

A corollary of a computational universe is the traditional notion of causality must be abandoned. Events don't cause other events, they compute them. Causality is a confusing metaphysical interpretation of science and not part of science itself. There are no variables of causality in any equation of science. It's an outmoded holdover from the old notion of a physical universe in which physical objects seemed to push other physical objects around to produce changes. But in Universal Reality where everything consists only of data being computed the usual interpretation of causality makes no sense and must be abandoned.

Because the universe isn't a physical structure causality loses its relevance. Since there isn't a single variable of causality in any equation of science whatsoever there is no loss to science at all in completely abandoning the concept. In an information universe the data states that are the actual reality of the universe are *computed* from prior data states. Nothing physical happens in a universe that consists of data. The elemental program that produces particle interactions computes them but doesn't cause them in any physical sense.

So causality is not really even part of science but a meta-principle or interpretation that just refers to the fact that events predictably precede other events in logical sequences. But this is because they are computed according to fixed rules, not because material objects somehow push other material objects around in predictable ways to cause them as was originally thought.

In a computational universe it's enough to know that programs predictably compute subsequent data states according to consistent logical rules. There is no necessity of any physical mechanism. Saying that computations cause events is meaningless. Adding 2 + 2 doesn't *cause* 4, though it does predictably *compute* a result of 4.

When causality is abandoned important new insights into reality and the nature of knowledge emerge. The main implication is that the history of the universe is not a causal history but a consistent history. This enables it to be understood as a single integrated information structure that is logically consistent in both temporal directions.

EMERGENCE

THE NATURE OF EMERGENCE

The universe directly computes only individual particle and particle component interactions through myriad applications of a single elemental program. However due to the exquisite design of the complete fine-tuning particle interactions in aggregate exhibit all sorts of complex *emergent structures and behaviors* that aren't initially expected from particle interactions themselves.

Thus everything in the observable universe above the elementary particle scale is emergent because it's a manifestation of particle interactions in aggregate rather than being directly computed by the classical scale equations science uses to describe them. Emergent programs include all processes above the particle scale, both inanimate and animate. Thus the programs of ourselves are emergent programs.

Emergence is based in the binding energies of particles and the balance of the four forces in atoms, molecules and the materials they form. Elementary particles combine to form compound particles, compound particles combine to produce the atoms of the elements, atoms combine to produce molecules, molecules join to produce chemicals, chemicals interact to produce materials, and materials interact to produce material objects and structures.

Some chemicals interact to form organic molecules and under the proper conditions these can eventually interact to produce single celled organisms, which can evolve over time to produce all the complex life of the universe including human beings. How life likely began is described in detail in *Universal Reality* (Owen, 2016).

So emergence is simply the way particles self-organize in aggregate due to their atomic and chemical reactions. Over vast stretches of time the self-organization has evolved into all the aggregate particle structures in the observable universe.

All such compound particulate structures that exhibit their own characteristic behaviors are emergent. Thus nearly everything in our familiar everyday world and the observable universe is emergent.

The emergent universe that exists today is the cumulative large-scale result of a single elemental program computing uncountable numbers of elementary particle interactions over the history of the observable universe. From myriads of these simple computations the entire emergent complexity of our amazing universe has blossomed. And how and why this beautiful and profound complexity is hidden within the complete fine-tuning of the quantum vacuum is perhaps the greatest mystery of all.

All the specifics of emergent structures and their behaviors are ultimately due to the particulars of the complete fine-tuning that determines how individual particles interact. Thus the entire spectrum of possible emergence is implicit in the complete fine-tuning from the beginning of the universe. It's truly a beautiful and awesome system.

The elemental computations of the quantum vacuum simultaneously compute all the mass-energy structures of the universe and the dimensional spacetime in which they seem to exist. This takes the form of the entanglement network of relationships among all elementary particles and their components including their dimensional relationships.

Thus the entanglement network is a single universal data structure incorporating every aspect of the observable universe. Simply put *the current state of the entanglement network is the observable universe* and it consists entirely of particle data in a continual process of recomputation.

Most all of science is concerned with emergent behavior. Thus the vast majority of the theories and equations of science are human *descriptions* of reality rather than human *discoveries* of equations that actually compute reality. How all the processes of the universe emerge from their particle interactions is the realm of all the sciences. Thus only the general principles of emergence can be outlined here from the computational perspective of Universal Reality.

This means that what can be called *reality mathematics*, the logic and mathematics the universe actually uses to compute itself, can be fairly simple. *Human mathematics* certainly shares the same logical fundamentals but most of the vast body of human mathematics are *invented* consequences of the common fundamentals rather than *discovered*. This is the solution to the long-standing question of whether human mathematicians invent or discover mathematics.

Thus there can be a single relatively simple elemental program that computes the entire universe in terms of its particle interactions. And most of the vast body of human science that describes the scientific universe is human discoveries of descriptions that emerge along with the phenomena they describe. This vastly simplifies the computational universe and Universal Reality as well.

This chapter explores the computational aspects of emergence and their implications for the fundamental nature of both reality, and for ourselves and our knowledge of reality.

OBSERVERS, THINGS, & EMERGENT DOMAINS

Because it's a computational system the universe can be considered a single universal running program that continually recomputes its data state.

This universal program consists of all the individual applications of the elemental program each computing a set of coherent entangled particles formed by an interaction event. Together in interaction all the separate applications continually recompute the entire data state of the observable universe in each current P-time tick.

The logico-mathematical consistency among the particle relationships produced by all these elemental applications is the current state of the entanglement network. The entanglement network is not a separately stored data structure but only the consistent logico-mathematical relationships among all the actually stored particle component values that make up the actual data of the observable universe.

Thus the entanglement network is the *overall view* of these relationships rather than an actually existent data structure. Since it's a view it's ultimately dependent on an observer recognizing it. Our description of the entanglement network is the view of a theoretical *omniscient observer*.

At the emergent scale the structure of the entanglement network can be described in terms of enormously complex hierarchies of interrelated sets of overlapping *data domains*. Data domains are aggregate areas of the entanglement network characterized by greater computational density and different computational types. In other words the universal aggregate of all the particles in the entanglement network automatically forms areas that look and behave differently from each other even though they are all ultimately connected in a single network.

Now it takes observers to recognize emergent domains. Because they are only recognizable at the aggregate scale, only observers whose simulations are tuned to recognize particular emergent structures are aware of them. Evolution has tuned observers to recognize meaningful emergent structures because this increases their fitness by giving them an understanding of how their environment functions at their own scales.

In general, through many stages along a spectrum, observers extract meaningful things and relationships from emergent domains to create dynamic simulations of their environments. They use these to plan appropriate actions in furtherance of their instinctual imperatives to help them function effectively.

Individual observers tend to identify individual 'things' on the basis of emergent domains, and they do this on an *ad hoc* basis. However things are artificial entities observers use to simplify their simulation models of reality because observers tend to think of things in terms of highly simplified conceptual descriptions and labels that ignore almost all the actual data structures of domains. Thus things are simplified mental representations of emergent domains meaningful to specific observers.

Thus the universe appears as a collection of meaningful data structures when viewed by observers. This is analogous to how patterns of individual colored pixels on a TV screen create meaningful moving images in aggregate to observers tuned to recognize them. Emergent things are organized patterns of particle interactions that are meaningful to observers tuned to recognize them.

To simplify their simulations of reality observers tend to think of their environments as consisting of discrete things and processes defined by sets of characteristics with relationships among them including interaction relationships.

However the actual structure of the entanglement network is continuous as everything is continually computed in interaction with other things so no perfectly discrete boundaries between things exist. All areas of the entanglement network have had some computational contact through the network and it appears there are no completely isolated computational areas because the very act of recognizing them would require particle interactions that would computationally connect them. Thus if any exist they are unknown.

The domains that arise in the entanglement network at the emergent level are hierarchical and often include sets of other domains and they may be members of sets of other domains. Thus the 'things' that humans tend to think in terms of are not well-defined discrete areas of the entanglement network but highly simplified extractions of data useful in the moment for the current thought process. Data domains are inherently fuzzy due to their connectedness through the entanglement network. They all are emergent aspects of a single interconnected universe of data.

As a simple example leaves, twigs, branches, trees, tree species and forests are all individual things humans might extract from overlapping hierarchical domains in the entanglement network on an *ad hoc* basis useful to some current thought process. Likewise surfers will think of a common ocean in terms of waves, oceanographers in terms of currents, and smelt in terms of tides. All these are 'things' isolated from overlapping data domains are of use in some current context to a specific observer organism. Ultimately all these overlapping domains are isolatable aspects of the single enormously complex domain of the entire entanglement network.

Thus discrete individual things as such don't exist in the actual universe above the particle scale. It's observers that identify and conceptually extract emergent things from the entanglement network. Individual organisms and species have evolved *perceptual and conceptual antennae* finely tuned to extract things useful to their own lives from the overlapping hierarchical domain structure of the entanglement network. Thus only observers are able to see the entanglement network in terms of its domains and extract meaningful 'things' from them.

Of all the 'things' we humans extract from the entanglement network the aggregate internal consistency of its dimensional relationships is one of the most important. We call this thing dimensional spacetime. Thus *dimensional spacetime is an emergent data structure extracted by living organisms rather than a pre-existing physical container for events.*

EMERGENT PROGRAMS

The aggregate applications of the elemental program that compute emergent domains can be considered emergent programs. Emergent programs are simply the aggregate computational interactions of all the particles in an emergent domain. Thus emergent programs are not separately stored code structures but meaningful aggregate computations of the elemental program that together compute emergent structures.

As with emergent things emergent programs are particular observer views of the computational processes of the observable universe. They enable the evolution of emergent domains to be usefully thought of in terms of *emergent programs* computing them.

Thus the entire observable universe can be thought of as a single running program consisting of the continual interactions of some particular *ad hoc* set of individual running programs which in turn emerge as meaningful aggregates of individual applications of the elemental program.

Everything in the observable universe is computed by executing organized patterns and sequences of a small set of fundamental operators in accord with a small set of virtual rules and data templates in the complete fine-tuning. But due to the exquisite tuning of the complete fine-tuning emergent programs manifest as independently meaningful programs that in interaction with each other compute the entire observable universe.

This is analogous to computer programs, which act as meaningful independent programs even though they are actually computed as organized sequences of *machine language operations*. It's the overall organization of machine language operations that gives a computer program a meaningful independent function at its emergent level. Thus the fundamental operators of the complete fine-tuning used by the elemental program to compute particle interactions can be considered the *machine language of the universe*.

However emergent programs are profoundly different than ordinary computer programs in several important respects. First *reality programs* can be defined as the programs that compute the actual universe as opposed to *computer programs* that run in electronic computers.

1. Emergent programs are meaningful sets of particle interactions

being computed simultaneously in every P-time tick. The only code involved is that of the elemental program that computes all the particle interactions individually. Thus all the computations of the universe can occur simultaneously in a logically consistent manner.

2. By contrast computer programs consist of *pre-written code sequences* that are executed sequentially to compute single data states at a time. Thus each computer program is effectively an independent prediction of the future states it will deterministically compute.

3. The deterministic futures of multiple computer programs can be inconsistent with each other because they run in separate computers. But reality programs all run in the same universe so they can't compute inconsistent futures. That would lead to logical inconsistencies in a computational universe that would tear the universe apart and as a result such a universe can't exist. If a computer program encounters an inconsistency it crashes and can be restarted but the universe can't crash and be restarted.

4. And how could such code ever be written and for how far into the future? It's simply an untenable model. This is why reality programs can only consist of simultaneous individual particle interactions all computed together in each P-time tick.

5. Also computer programs are written by programmers to fulfill specific pre-determined functions but emergent programs have stochastically evolved over the history of the universe on the basis of relative fitness. Reality programs are self-organizing evolutionary processes selected among the fittest aggregate groupings of elemental computations.

6. There is no general process that writes higher-level code strings to compute all emergent processes. However there are exceptions to this rule. DNA is prewritten code that computes the development and operation of organisms, simulations contain programs to compute the actions of biological organisms, and of course computer programs are pre-written emergent programs as well. These are valid examples of actual *higher-level programs.*

7. Lastly computer programs compute data in computer memories, but reality programs compute the data of real actual things in the substrate of reality. Of course computers are real things that exist in reality as well. So everything including computer programs is ultimately computed as a reality program.

8. All the data in the universe is the data of the real actual thing it is in the universe. The reality of the data of a computer program that simulates the universe is its bits and bytes in computer memory. The reality of the data of that computer program running in a

computer is the real computer program running in the computer. The reality of the data of the actual universe is the actual universe. The reality of a computer simulation of a real universe is a computer simulation of a real universe. So everything that exists is the complete data of what it actually is and the running program that computes it, and even higher-level programs are ultimately the reality programs of themselves. The emergent program of DNA that computes an organism is itself being computed as a reality program by the universe itself.

The universe consists of continual simultaneous recomputations of all data states rather than sequences of static data states. A current static data state of the universe or any of its programs exists only as an observer snapshot of a continuous process. Observer simulations of reality tend to consist of sequences of static data states for easy comprehension, while reality itself consists of a continually flowing process of running programs. Thus it's emergent programs that are fundamental rather than the emergent data states they are continually recomputing.

Thus everything we see around us, including ourselves, is every bit of its data in an active process of computational happening. Everything that exists, including ourselves, exists as data within the quantum vacuum of reality where we are all continually recomputed together into continuing existence in the current present moment.

Thus the universe can be thought of as a single universal program that can be understood in terms of myriads of individual emergent programs each continually recomputing its processes in interaction with the others of its computational environment. Thus, at the emergent level, the single universal program can be thought of in terms of individual programs that continually recompute the entire data state of the universe in interaction with each other.

However just as there is no single set of discrete emergent things there is no single set of discrete individual programs. Individual programs compute the data states of 'things' so they too are defined on an *ad hoc* basis by observers to explain how the things they think in terms of function. Even so this is clearly a meaningful aspect of reality as it's essential for the successful functioning of organisms to view the world in this highly simplified manner.

Thus emergent programs exhibit the appearance of independent programs with independent functionality and even in the case of biological programs of purposefulness while ultimately remaining organized simultaneous interactions of elementary particles in aggregate.

THE LOGIC OF THINGS

Emergent laws are laws that *describe* emergent processes but don't actually *compute* them. All the laws and structure of the universe save for the elemental program that computes particle computations are emergent.

The emergent laws that describe the states and processes of the universe are those of the experimentally verified laws of science insofar as they are known. Universal Reality accepts all the established theories of science, always subject to revision based on new evidence. In this way our theory maintains complete consistency with the actual equations and logic of science even though our *interpretations* are new and improved.

The emergent laws of the observable universe contain a deep structure called the *logic of things.* Humans and other organisms use these basic principles to understand and function within their environments, and the deep structure of the laws of science is based on them. The logic of things is the fundamental logical rules that underlie the interactions of everyday things at the emergent classical level of our simulation of reality.

These fundamental rules emerge naturally with the processes they describe and are extracted and codified in our simulations of reality. They are the basic logical rules that biological organisms use to make sense of their surroundings and function effectively within them. Because it's also used to design and control robotic systems the logic of things is becoming clearer as robotic functionality increases.

The logic of things is based in a nexus of simple classical rules such as single things can't be at more than one place at a time. To get from one place to another things must move along some path between them. Things don't arbitrarily appear and disappear. Existent things must be somewhere. Events have causes, and so forth. The logic of things is the complete set of these simple fundamental rules of our simulation's model of the world of things.

These emergent laws are in approximate descriptions of the fundamental logical rules by which emergent programs operate. They ultimately derive from rules intrinsic to the complete fine-tuning. They tend to have exceptions and inherent limits to their accuracy especially with respect to smaller and smaller aggregates. Many stochastic processes tend to converge on exact results in aggregate as demonstrated by half-lives and the laws of gases and the logic of things describes the emergent behavior of these aggregates.

It's in the nature of the *super-consistency* of the universe that the emergent laws of nature are largely inter-consistent at whatever level they apply. Emergent laws emerge along with the aggregate processes they describe and are discovered by humans who model them as best they can in their simulations of things and their interactions both in their own lives and in their sciences. Certainly the overall logic of things emerges with the processes themselves but it also reflects the intrinsic structure of how simulations model those processes.

Thus the logic of things describes emergent reality as it's simulated by particular organisms however the basic rules must be similar across all organisms because they all must accurately map the actual logic of reality to successfully function within their environments.

INANIMATE PROGRAMS

The inanimate programs of the universe compose a vast hierarchy of overlapping systems from the largest scale cosmological processes down through the geological processes of individual planets, to the building, interactions and erosion of the smallest grains of minerals, and all the flows of energy and materials that connect them. These are all integral aspects of the universal program that can be isolated and studied on any number of individual bases.

These processes are described in wonderful detail by the physical sciences and new discoveries are continually adding to our knowledge of how they operate, interact, and have evolved into the universe we observe today. All that Universal Reality adds to this picture is to reveal they are emergent computational processes that are best viewed as running programs rather than physical processes. They are all emergent manifestations of vast numbers of elemental computations that have been

124

selected over the history of the universe through their collective fitness within their environments.

The inanimate programs of the universe are characterized by being non-purposeful. They operate on the basis of immutable stochastic laws of nature at the particle level. In aggregate these laws determine which programs tend to emerge and persist within the total environment of programs through at any particular time and place within countless generations of interactive computations. The universe we observe today is the current result of vast numbers of quasi-random computational interactions over its entire history. It's the current manifestation of the complete fine-tuning that encodes its fundamental design.

When it comes to specifying the particular programs that constitute the inanimate universe observers have great latitude. They can think in terms of the individual chemical reactions that form rocks and minerals, the deposition and erosion of geological processes, the plate tectonics that drives the building of mountains and the movement of continents, or the cosmic scale processes that create stars, planets and galaxies, or anything in between.

All these views are legitimate and all identify computational processes that can be studied as the actions and interactions of individual programs within the complex overlapping hierarchy of the whole. This is true of all the processes of the universe. All are parts of a single universal program but all subsets can be meaningfully viewed as individual programs at all levels of the universal entanglement network of domains.

In general the individual programs of interest to observers tend to be identified on the natural basis of computational domains but there is no theoretical limit on what set of computational processes could be considered an emergent program. It all depends on how individual observers choose to extract them from the entanglement network.

Inanimate emergent programs can be thought of as computing all the non-purposeful processes of the entire observable universe. Over time in favorable environments inanimate programs have gradually evolved the purposeful emergent programs of living organisms. There is a vast continuous spectrum of inanimate through minimally purposeful through highly intelligent emergent programs that reflects this ongoing evolution.

LIVING PROGRAMS

Inanimate programs are the direct emergent manifestations of their elemental particle interactions but *biological or living programs* have evolved to purposefully compute their interactions with the other programs that constitute their environments.

They do this on the basis of internal data models or *simulations* of themselves within their environments that enable them to model possible future actions and valuate and select among them on the basis of instinctual imperatives. This enables them to further their survival and the survival of their group and ultimately of their species.

Thus living programs are characterized by having internal models of themselves within their environments that enable them to act purposefully within their environments. In this sense living programs include all biological organisms including humans and could potentially include artificially intelligent robotic organisms as well.

Even inanimate programs can be said to experience each other in the changes their interactions produce in their own data. *Simulations* just take this fundamental mechanism another step by constructing internal data models of things they interact with and experiencing them more explicitly as changes to their data representations.

This allows an organism to experience things as changes to its internal representation of them. Simulations are subroutines within the programs of living organisms that give them the ability to purposefully compute their actions by simulating them internally before acting. Thus simulations act as purposeful operational control systems for living programs. Simulations are described in detail in the next chapter.

The fundamental nature of all emergent programs is emergent code. Thus we are the complete program code of ourselves running in the present moment as P-time ticks. The vast majority of this code is continually computing the processes of our somatic bodies and all the complex homeostatic systems that operate and maintain them. At the highest level our simulation is an add-on that enables us to act purposefully within our environment. Thus 'we' are the program of our entire being including all these systems functioning together to operate our entire being purposefully within the world.

As a result our whole program functions as a finely tuned living organism even though it's ultimately all being computed at the 'machine language' level of particle interactions. It's the simple fact that the elementary particles that compose us are organized into the biological structures of our bodies that enables their interactions at the particle level to manifest as the emergent computational systems of ourselves.

Our own programs continually interact with the other running programs that make up our environments, and the results of all those computational interactions generate the facts of our lives and our effects on our world. It's all an enormously complex computational process in accordance with the laws of nature including those of human behavior.

In effect we are the biological robots of ourselves. Our programs are living, intelligent, sentient, conscious, emotive, partially free-willed, purposeful and self-modifying and we operate effectively and autonomously within our environments. Our programs are a wonder of design that has evolved stochastically over eons in interaction with the programs of our environments, and our programs are even capable of reproducing their kind.

We are the fully human, partially free willed, living conscious robots of ourselves, and our true nature is that we are the running programs of ourselves rather than physical beings. As beings consisting solely of the information of ourselves being continually recomputed by our emergent programs we inhabit an information universe rather than the illusory physical universe our simulation represents it as.

This entire program is the true nature of what and who we really are. We are not just a consciousness carried along by a physical body, we are the entire program of ourselves down through the hierarchy of all our systems to its most elemental operations at the particle level, and all this is ultimately only running code continually recomputing our data.

Though initially counter intuitive this is really not so difficult to understand and can even be directly experienced if we put our mind to it. All we have to understand is that the *information* of all the processes of our bodies of our being down to the finest level is all that's actually observable in any way whatsoever. Even our apparent physicality is ultimately separable into the *information components* our simulation labels as physical in combination.

So just put all the information of our apparent physicality together including the complete information of all the actual particles leaving nothing out and this complete information structure is all that we actually are. It's always what we have been and it's just a matter of recognizing this verifiable fact. Our awareness of the internal processes of ourselves is simply our experience of our own program running within us even as it reads and comprehends this sentence.

So the recognition that we are our information, the information of our complete running program, in no way diminishes us as humans. It doesn't change us in the least; we remain fully human exactly as we always were. We think, feel, and act exactly as we did before. There is now just a deeper understanding of what we really are compatible with the deeper understanding of the entire computational universe of information revealed by Universal Reality.

The interactions of emergent programs with other programs are all computational and they all consist of information only. For example the program of a human being hit by a bus exists only as the information generated by the computational interaction of their programs. But this information representing the breaking of bones or loss of life is the reality of the actual event in the real world. Everything is computational, everything is information, but the complete information of things is the actual things because it exists in the substrate of reality itself. The complete information structures of things are the things themselves.

Because aggregates of particle interactions naturally form such rich emergent structures thanks to the complete fine-tuning, emergent programs emerge as discernible running programs that can be independently named and described in terms of their overall function. All the data of what things are continually interacts to compute the function of the whole. And the function of the whole is determined by the computational interactions of all the programs that run within it.

FREE WILL

All true randomness is quantum randomness and occurs only at the particle level as dimensional indeterminacy is computed. All the apparent randomness and freedom of classical level events and actions is a structural amplification of quantum randomness up to the emergent level (or in many cases simply the non-computability of extreme complexity which is not actual randomness). The non-predictability of turbulent flows, the weather and other extremely complex phenomena are combinations of both effects.

128

Thus all emergent programs can be said to have some degree of free will in the sense that their behaviors aren't completely determined by their external environments and this includes all living programs. So what we call our free will is ultimately an amplification of the indeterminacy of our quantum processes up to the level of operational decision-making.

So even though the universal program is computational it's not completely deterministic because its computations incorporate a constrained degree of stochastic randomness at the quantum level. Therefore the universal program, and the emergent programs that run within it, exhibit varying degrees of freedom from deterministic causation depending on their individual information structures. In living organisms this is the ultimate basis of free will.

Many but not all processes of the universe incorporate quantum randomness. For example the fundamental conservation laws are exact, at least to the level of granularity of elemental reality, but computations involving dimensionality generally exhibit some degree of quantum randomness due to the random oscillations in the processor as it computes space and time velocities as explained in the chapter on *Quantum Reality*. If this model is correct then all randomness and as a result all free will originates in the random oscillations between space and time velocities as the universe is computed by applications of the universal processor.

While all the processes of nature obey the laws of nature, they do so in a non-deterministic manner so their evolutions are never predictable below certain levels of detail depending on their individual structures. We know mountains will erode over time in a generally predictable manner but it's impossible to know exactly how this process will play out down to the individual grains of stone.

The amount of freedom the programs of various systems exhibit is highly dependent on their internal data structures and whether they tend to concentrate or damp out quantum randomness at emergent scales. For example human mechanisms such as digital clocks and industrial robots are designed to almost completely eliminate the expression of randomness at the level of their design operations, while others such as dice or lottery drawings are designed to maximize it. The number of quantum processes may be the same in each but whether their randomness is magnified or damped depends on their design.

Living organisms including humans are designed to act purposefully in response to external stimuli but to exhibit a significant degree of freedom in doing so. We respond meaningfully to external stimuli but our actions are not completely determined by our environments. Fundamentally this free will is due to the presence of quantum randomness within the elemental processes of our programs. We are designed to make individual decisions based on relative evaluations and weightings of masses of information bubbling up the hierarchies of our programs from below. Thus minor changes either in the emerging information, or computational choices among information streams, can effectively magnify the degree of effective free will exhibited.

Since the actual computational interactions of complex programs like humans with the programs of their environments ultimately take place at the level of elementary particles, our emergent level processes are considerably insulated from the emergent level processes of our environments. So it's our hierarchical complexity and structural design that allows humans to exhibit considerable free will. Thus randomness is damped in some areas to maintain structure and function but enhanced in other areas to test novel adaptations to changing environmental situations that enable more effective responses to be developed.

Thus the randomness in the elemental computations of a biological organism is ultimately the source of its free will but the effect is focused by the hierarchical structures of its decision-making processes.

Almost all our decisions are computed at the unconscious level, and almost all are computed with a modicum of free will. Consciousness, in its quality control function, allows a small additional degree of free will to modify or override unconscious decisions.

As most humans mistakenly identify with their consciousness, the relatively minor conscious decision making capacity is what most people think of as 'their' free will because they consciously experience it as such. This explains why people tend to think of their free will as the freedom of their conscious self to override their *own* instinctual imperatives, but that's a very minor part of the whole decision making process since almost all our decisions continue to be made at the unconscious level.

Thus free will is an intrinsic attribute of all emergent systems depending on their design because their actions are never completely determined down to the particle interaction level. This means that artificial intelligence and robotic systems can certainly be designed to exhibit free will. The purposeful structure of the simulation is designed to do just that while generally remaining within the bounds of the instinctual imperatives to avoid self-harm.

Ultimately the amount of free will we have depends on the number and richness of available choices we can imagine and choose among. This in turn depends on the robustness of our simulation. This varies greatly from species to species and from individual to individual. Thus of two people in the same situation one could see several reasonable choices while the other might see only one and effectively have no free will at all in making it.

THE GENERAL PRINCIPLE OF EVOLUTION

[Note: *data, forms* and *information* are used here more or less interchangeably depending on context. Data are the fundamental elements of the observable universe. Information is simply meaningful data and implies an observer or observers that meaning is to. Forms refer to the *structures* of data or information, whereas data or information emphasizes their *content*, but since content and structure are actually the same when it comes to data the usage is mainly a matter of context while the meaning is essentially identical.]

Emergent forms and the programs that compute them continually appear, interact, transform, and disappear. In effect all the emergent forms of the observable universe are in a continual *computational competition* in which some persist and multiply and others fade and vanish.

All emergent forms remain the same unless they are changed computationally. This is effectively an evolutionary law of inertia. All things stay the same unless they are computationally changed and when things do change they change only as the result of computations.

Emergent programs evolve over time in interaction with the others that form their environment according to a simple fundamental principle that can be called the *General Principle of Evolution* (GPE). This principle simply states that those forms or programs that win computational interactions with other forms are those that persist at the expense of others.

Though this principle seems trivial on its surface it's actually quite profound because it's the fundamental rule that determines how the entire universe of emergent forms and programs evolves. The GPE is an explanatory rather than a computational principle. It merely describes how the universe of individual forms evolves. It doesn't compute the evolution it just describes its emergent level results.

Darwinian evolution is simply a special case of the GPE that applies to biological programs that reproduce their kind. The survival and procreation of individuals of all species is selected via their computational interactions with their local environments including individuals of the same and other species. Species whose individuals prosper relative to others are said to have increased fitness within their environments, and populations of these species tend to increase over time because their individuals do. In this way the mix of individuals of all species becomes better adapted to its common environment.

[Actually most evolutionary selection is a matter of chance rather than fitness but on average over many events it's relative fitness that slowly guides evolutionary change.]

Thus the GPE simply states that all programs can be considered to be in computational competition with others and those best at surviving their interactions tend to persist and spread. This principle plus the details of the complete fine-tuning are sufficient to explain the classical scale evolution of all the emergent programs in the universe including ourselves.

In general as forms interact they continually transform into other forms. The entanglement network that contains them all continually evolves and its domains continually change, combine and dissolve into other domains. Thus the emergent programs and things observers base on domains do likewise.

So the survival of individual forms depends on how well and for how long they survive their computational interactions with other forms. Thus the overall evolution of the forms and programs of the observable universe depends on what forms tend to emerge over time in interaction with others. On average those that survive short term tend to survive longer term but everything depends on details of the individual interactions.

For biological species that reproduce their kind, their species tends to survive and multiply at the expense of other species if they are adaptive in their interactions within their environment of other animate and inanimate programs. Thus Darwinian evolution is a simple self-evident description of the progressive results of computational interactions.

Thus the evolution of the observable universe consists of the continual computational interaction of all its individual programs. This process continually selects the mix of programs that constitutes the observable universe at every present moment. The complete fine-tuning of the quantum vacuum manifests its hidden design through time through the interactive evolution of all its emergent processes.

Thus the programs that exist in the observable universe tend to converge towards the most successful and adaptive mix at any given time. In this way the observable universe tends towards expressing itself in the optimal possible manner through an automatic evolutionary process. The observable universe automatically evolves towards and converges on the *fittest mix* of running programs.

However due to the lags in computational results spreading through the system this mix continually changes over time. As computational changes spread through the system the programs that constitute the environments of other programs change and individual programs must adapt to their new computational environments or be superseded by fitter programs. Thus it's the slow spread of computational changes across the whole system that allows the universe to continue to evolve through time rather than permanently reaching a stable end state.

Through the individual interactive selection of programs the entire mix of all programs, both biological and inanimate, becomes better adapted to the universal environment of programs. Local programs become better adapted to each other. However this is a never-ending process due to the time lag of computational changes propagating through the entanglement network. This time lag concurrently slows the increase in entropy in a dynamic system like the biosphere.

So the GPE is simply a new way of understanding the evolution of our universe rather than any difference in the way evolution works. Thus the gradual evolution of the universe from the big bang through the origin and development of life including ourselves remains largely as science describes it. It's just viewed in terms of computational processes or programs rather than biological species exclusively.

There is another important point to be made. Critics of Darwinian evolution rightly point out that random combinations of molecules would never be sufficient to create the vast range of species that has existed. But that's not how evolution works.

The evolution of emergent forms including biological forms is progressive rather than randomly restarting from scratch at every step. Each step builds on previous steps. At each step emergent building blocks are produced that are useful in constructing higher-level emergent programs. Thus *meiosis* typically results in the rearrangement of tested building blocks that in new combinations are likely to form new forms of viable life. And even individual *mutations* sometimes act similarly, and those that don't are unlikely to be reproduced.

So instead of randomly rearranging individual molecules to create new species, evolution can simply rearrange sets of already tested building blocks of life. The Hox genes that control the general arrangement of appendages along bodies depending on the sequence of their expression are a good example (Wikipedia, *Hox gene*).

By analogy evolution works by testing new arrangements of Lego blocks rather than new arrangements of molecules of plastic. The fact that such building blocks tend to emerge naturally from particle interactions and be used to evolve ever higher-level emergent programs is part of the wonderful mystery of the complete fine-tuning.

HUMAN COMPETITIVENESS

Like all programs human programs compete for survival in accordance with the General principle of evolution. However human programs compete purposefully since their actions are driven by the instinctual imperatives of individual survival and reproduction. The extreme effectiveness of human competitiveness is precisely why humans have spread across the planet and gained so much control over it. Thus it has been strongly selected for in the human species.

Over roughly the past 10,000 years humans have become by far the most successful species on the planet in terms of their ability to purposefully affect it. This is due to a number of factors primarily the ability to construct tools and technologies to manipulate the environment, and the ability to store and disseminate knowledge. However these abilities simply augment the fundamental driver of human success, which is the capacity for ruthless competition over resources.

In the continual struggle for personal and group survival those who competed more effectively have tended to survive at the expense of the less competitive. Thus the instinctual imperative for ruthless competitiveness is strongly carried by those that did survive. Thus modern humans are here precisely because we carry the instinctual imperative to survive whatever it takes. This has become an intrinsic part of our human nature.

The history of human civilization is largely the history of the complex details of the competition for resources between individuals and human groups and other species. Of course the competitive instinct can be damped down by cultural rules especially in times of relative plenty, but it's always there ready to burst forth in potentially lethal competition between individual humans, human groups, and between humans and other species as they compete for the necessary resources to survive and prosper.

Today as human population and technological power continue to soar as sustainable resources continue to dwindle the very instinctual imperatives that enabled our global success now threaten to destroy us and the biosphere upon which we depend. Human competitive success has led to human overpopulation to the point of depletion of the sustainable resources upon which our species and much of the biosphere depends.

In the future as sustainable resources continue to dwindle and human overpopulation continues to increase there will inevitably be greatly increasing competition for fewer and fewer resources. Given the ruthless competitive instincts that got us here competition is likely to increasingly explode in potentially lethal conflicts between individual humans, human groups, and between humans and other species as all compete for the necessary resources to survive and establish competitive advantage.

Thus it seems inevitable that greater and greater conflicts in the form of wars and interpersonal strife will arise until human overpopulation comes into balance with dwindling resources. Human instinctual nature is more or less the same everywhere. The worse case scenarios we see throughout history and currently see elsewhere can just as easily happen anywhere and everywhere given conditions of want.

Since the instinctual imperative of ruthless competitiveness in humans cannot be changed the only solution is a new system of enlightened government to transform it into a universal human competition to produce the richest and most viable sustainable biosphere for the good of all humans and other species possible. This will hopefully be explored in an upcoming book.

HIGH-LEVEL EMERGENCE

Individual emergent programs are all stochastic variants on themes implicit in the specifics of the complete fine-tuning. And the evolution of the entire observable universe is a random variation on a single underlying theme that was present from the beginning. This is true of all the classical level programs of the universe both living and inanimate.

As with inanimate programs the programs of biological life forms can be considered in terms of individual organisms, social groups, or even in terms of the historical processes of families, nations and cultures. All can be viewed as the operation and evolution of overlapping hierarchies of individual programs within the context of the universal program

There is no apparent upper limit to emergence. The flows of individual lives from birth to death, the great flows of civilizations and history and evolution, the interactions of all the programs that constitute the biosphere and total ecosystem of earth (Lovelock, 1995), and no doubt even greater as yet undiscovered processes, can all be considered as interacting programs emerging at every level from the interactions of their constituent programs, and all ultimately from structured sequences of elemental operations. Computer analysis of big data is progressively revealing many of these hidden computational processes and how they function as high-level emergent programs.

All these individual programs are computationally intertwined processes that are part of the single running program of the observable universe. Each can be teased apart into individual programs running in the context of all the others at any level. All these programs are the progressive emergent manifestations of the innumerable ongoing computations of all the particle components in the universe.

All individual organisms are enormously complex hierarchical programs, but the much more complex computational *interactions* of these individual programs determine all the great processes of human history. Social interactions among biological programs are programs as well, and are part of the running program of the evolution of the earth and all its individual life forms, natural cycles, and geological history and the enormously larger cosmological programs that have created and maintain our universe.

Insect swarms, large flocks of birds, migrating herds and colonies of eusocial species are all examples of the emergent phenomenon of collective intelligence as they routinely exhibit problem solving behaviors lacking in their individual members. Emergence also accounts for the collective intelligence of neurons in brains, which exhibit intelligence in aggregate though no single neuron can be considered intelligent. Likewise single machine language operations in a computer have little meaning but organized patterns of these operations can compute just about anything.

Just as a bacterium might well be aware of the firing of a single proximate human neuron but would have no possible comprehension of the workings of the human mind that neuron was part of, we humans are aware of many of the individual processes of our universe, but likely have no idea at all of the hidden deep scale processes they manifest at higher levels of emergence, much less any inkling of emergence at the universal scale.

Due to the incredibly rich and profound design of the complete fine-tuning that informs the process of emergence, the entanglement network exhibits *super-consistency* at all levels of emergence. The entanglement network manifests as a vast super-consistent hierarchy of emergent programs each consisting of aggregates of lower level programs.

The entire structure of emergence is super-consistent in the sense that every individual process is self-consistent, every level is internally self-consistent, and every level is logically consistent with every other level. The entire information structure of the observable universe is a computationally super-consistent logico-mathematical entity. The descriptive laws of nature that emerge at every level are themselves consistent with the laws of all other levels because each derives from those below. The entire hierarchical unity of the universal program is a single multiply self-consistent computational process at every level.

[Note that individual observer simulations though part of the entanglement network need not be entirely consistent with each other or even internally self consistent because they are simplified *ad hoc* mappings of a consistent entanglement network used by organisms to heuristically compute their actions. Thus it's not unusual for humans to hold wildly inconsistent beliefs even though their overall programs are computed consistently within the entanglement network.]

Thus at every level there are certainly as yet undiscovered laws in operation describing as yet undiscovered programs running hidden within them. And the continuing discovery of these hidden programs will certainly lead to a much greater understanding of how the universe and life on earth functions, and a greatly increased ability to predict and benefit from this knowledge that could potentially guide future developments in a beneficial manner.

By adopting a systems approach to the understanding of the great processes that guide our universe, the evolution of our earth, and our individual lives, we improve our ability to accurately simulate these processes and thereby guide our progress into the future in a more effective manner. Whether this will be done for the benefit of those who control such knowledge or the broader benefit of civilization and the earth as a whole remains to be seen.

The very fact that the universe of remarkably complex and meaningful programs emerges naturally from seemingly blind inanimate processes suggests there could be something more going on here. That somehow there could possibly be some sort of feedback mechanism from the strongly meaningful higher-level laws that somehow tunes the complete fine-tuning that produces it. What this mechanism might be if it even exists is the ultimate mystery, and hopefully the relation between the complete fine-tuning and emergent meaningfulness at the universal scale can be further clarified in the future.

Thus considered all together over the life of the universe we observe its evolution at least locally towards higher and higher levels of emergence and complexity exhibiting an intrinsic though often unrecognized evolution based intelligence implicit in the complete fine-tuning of our universe. There is certainly much more to be discovered here though the limits of human intelligence may ultimately limit human though not necessarily non-human understanding.

CONVERGENT EMERGENCE

The evolution of the universe is nondeterministic due to the inherent randomness of quantum processes, but it's strongly constrained by the rules of the complete fine-tuning. Thus the evolved history of the observable universe is one of many possible variations on a single grand design that existed at least since the big bang.

The precise specifics of the complete fine-tuning determine the general direction of the evolution of the observable universe and constrain its deviations. They deterministically produce the laws of chemistry and other elemental laws but they also stochastically determine the actual chemical reactions and evolution of the large-scale structure of the universe.

Thus emergence inevitably converges towards ends already implicit in the complete fine-tuning as the universe was created. Evolution is a process of *convergent emergence* along random paths towards statistically uncertain but largely predetermined ends.

Thus the general design of the universe and the laws that describe it were predetermined at the big bang, but the myriads of individual details of that design are the result of quantum randomness deriving largely from the creation of dimensionality in the entanglement network.

For example along with the observed general large-scale structure of the cosmos the fixed laws of chemistry seem to naturally evolve some form or forms of intelligent life where conditions are favorable. More intelligent programs tend to be more successful over long periods of interaction and will almost certainly acquire certain characteristics. They will likely have appendages able to manipulate their environments to create supportive technologies. They will tend to develop means to record and share collective knowledge. And they will likely be aggressively competitive and prone to ruthless violence to promote their success both individually and as a species.

To what extent this initial stage of highly intelligent technological life can be superseded by a more benign, wiser and compassionate intelligence is unclear. Ultimately that would be adaptive as it would foster the longer-term survival of advanced civilizations but that could be at the expense of thousands of 'intelligent' civilizations that destroyed themselves and their planets before reaching that stage. The jury is still out based on the conflicting human evidence.

In any case it seems likely the universe inevitably tends to evolve towards becoming aware of itself through the senses and intelligences of the life forms it naturally produces. One could say the evolution of the universe is an evolution from unconsciousness to self-consciousness, a process of producing sense organs, sentience, intelligence, and consciousness through which the universe becomes self aware.

This evolution has occurred in profusion on earth in the form of the vast interconnected network of the individual sensory organs and intelligences of humans and all other species that continually exchange incredible volumes of information about the world. And it's greatly extended in the collective intelligence and knowledge we humans store outside our brains in various media including the Internet. And an enormous range of scientific technologies from the invention of telescopes and magnifying lenses to remote sensing and supercomputer visualization has exponentially enhanced our sensory abilities.

This great explosion of sentience, consciousness and intelligence on earth effectively makes our small planet the brain of the universe, at least the only one of which we are currently aware. And this brain enables the universe to become consciously aware of itself in exceptional detail. Every thought we think, every feeling we feel, is the universe thinking and feeling itself through us, and this is equally true of the sensations and intelligences of all the other living creatures on our planet. Every one of them is the universe experiencing itself and knowing itself through the life it has evolved and the technologies that life has produced.

Thus evolution can be seen as the universe gradually awakening to its own existence and to some extent gaining conscious control of its destiny. While previously it was the presumably blind complete fine-tuning that programmed the universal program through a long slow evolutionary process, the universe now suddenly gains the ability to begin to consciously program itself. Who knows to what extent this capability may develop? In any case we humans now act as the apparent main repository of the universe's self-conscious intelligence as we look into the future and begin to guide it towards its ultimate destiny.

The ultimate destination of convergent emergence is of course speculative but it's very likely there is one that may already have been implicit when the universe began. The entire evolution of the universe may be only the process of making this originally implicit virtual design actual. Whether it's an eternal dark entropy death in which nothing more ever happens, or some sort of truly cosmological intelligence that ultimately fills the entire universe with awakened consciousness is unclear. But it certainly seems possible that the entire emergent system produced by all the elemental computations in the universe could already be manifesting as some sort of cosmic mind in the act of awakening.

If the aggregate effect of all the individual firings of the neurons in the human brain emergently manifests as the intelligent consciousness of a human mind, then perhaps the aggregate emergent effect of all the sensory organs and minds of all living beings on earth also functions as a super organism. Perhaps even all the elemental quantum computations in the universe manifests as some sort of cosmic mind, or at the very least something far beyond our current understanding. It certainly manifests as the observable universe including the minds of all its living beings. And at the highest emergent levels there is likely much more going on far beyond the comprehension of the human mind.

It's the continual computations of the quantum vacuum that give us our own individual lives. The universe is not a biological organism but it is a living computational organism because it functions autonomously with no external mover. Every one of us is an integral part of it in our own ways according to our own forms. To what extent its ultimate destiny is to become fully conscious and purposeful is unclear but certainly intriguing.

THE INTELLIGENCE OF DESIGN

The total intelligence of design encoded in the universe is awesome and immense. Though certainly the product of evolutionary processes based in the design of the complete fine-tuning rather than the product of any designer god or alien programmer it's innumerable orders of magnitude greater than any possible human intelligence.

After all it has produced and continues to produce the enormously intelligent design of every last entity in the universe as part of a single intelligently integrated computational system. By contrast we humans are incapable of designing the simplest component of even the simplest living organism from scratch, and even if we could design it we would still face the impossible task of constructing it. All that we are able to design we accomplish only by tinkering with what the universe has already designed.

Thus the total intelligence incorporated into the entire universe is far beyond the comprehension of human intelligence and no doubt will always remain so. Even though the intelligence of design of the universe is far beyond even that of any god we humans could possibly imagine the important point is that it raises the universe itself to god-like status. Thus if we want a god, the universe itself is the only viable candidate. If anything is divine and worthy of worship, it's certainly the awesome living intelligent design and happening of the universe itself.

THE SIMULATION

WE LIVE IN A SIMULATION

The world we experience around us is an illusion. It's a computational *simulation* of reality rather than actual reality. The true nature of everything in this world is not as it appears to be including even our own true natures.

However the simulation we experience our entire lives within isn't coded by any devious alien programmer but by own brain. The mind of every living organism continually computes its own dynamic simulation of reality and every individual organism believes the simulation its mind computes is the real actual world in which it lives.

[The popular hypothesis we live in a simulation programmed by other beings is almost certainly false. There must always be a real more complex universe within which any simulation is programmed and maintained. Thus any simulation must exist within an actual reality that supports it. The evidence that reality must be computational is overwhelming, and simulations are computational by definition. Any computational reality must be logically consistent and complete or it would tear itself apart at the inconsistencies or halt at the incompletenesses. This means it must exist as a single interconnected information network. This implies that all parts must be ultimately knowable by tracing the connections in its information network.

Thus an observer with sufficient intelligence should always be able to discover the actual reality that supports his simulation by tracing the information network beginning from his location within it. Thus though it's clearly possible for some beings living in a simulation to be programmed not to discover it, it's also overwhelmingly likely that beings with as much scientific knowledge and as intelligent as humans would have been able to discover they lived in a simulation if they actually did.

Thus the lack of any evidence we live in a simulation is overwhelming evidence that we don't. The usual argument that it's easier to construct a simulation than a real universe ignores the fact that a real even more complex universe is necessary to support any simulation that exists within it.

This is also debunked by the enormously greater complexity of our universe than a simulation would need to have which is absolutely necessary to accept as long as we believe in the logical consistency of science. The pixels of a projected reality just wouldn't operate by the rules the emergent processes of our universe operate by. And why would the programmers go to the enormous effort of programming the entire observable universe if they didn't have to? So the proposition we live in a simulation because it's easier to construct is belied by the fact that the universe we live in is demonstrably not simple.]

Every organism thinks its simulation is the true reality it exists within but this obviously can't be true because every individual's simulation is different. My simulation is much different than yours, and even more different than that of a fox, a shark, or a beetle, but all are equally valid for the species concerned and each individual of every species believes its simulation is the true actual world in which all other species live. But clearly this is an illusion.

On the other hand simulations aren't completely imaginary worlds, they are internal mental models of the real actual world in which all organisms live. They must have sufficient logical correspondence with the actual world they simulate to enable an organism to function effectively within the actual world.

Thus simulations are very simplified *logical mappings* of the changing structures and events of the actual world that are extensively ornamented with appearances, meanings, and valuations added by an organism's own mind.

So there is sufficient logical correspondence between the simulation and the actual world to enable the organism to function effectively within it, but all the *appearances* the simulation presents to the organism are unique and specific to the organism because they are products of the *interaction* of the organism with the world rather than attributes of reality itself.

Since the actual world is entirely a computational process consisting of programs computing their data clearly it doesn't have any appearance at all. Thus all the appearances of the bright world around us are constructed by our simulation on the basis of our perceptual interactions with it and then presented to us as actual attributes of the world even though they are actually produced in our own brains.

Since actual reality includes everything that exists, both we and our simulation are clearly part of reality. However we experience actual reality only through our simulation of it. Other than that we have no knowledge or experience of actual reality at all. Thus our simulation continually stands between us and actual reality and we experience actual reality only through our simulation of it.

The fact that the reality we experience exists entirely as an information construct in our own brain is extensively confirmed by cognitive science, however the implications for the nature of reality never seem to be taken to their obvious logical conclusions. However they are fully explored in this chapter and they reveal and confirm the computational nature of reality.

HOW THE SIMULATION WORKS

Simulations have evolved to enable living organisms to function more effectively within their environments. Thus they take many different forms among different organisms and even in different individuals. However their basic functional structure and operation are all remarkably similar as they must be to enable all organisms to function in the same actual world.

Simulations are programs that construct and continually update an internal data model of the organism within its environment on the basis of which the organism computes its actions. This information of self within environment enables an organism to intelligently compute its actions in furtherance of its instinctual imperatives of survival and procreation. Thus living organisms are able to demonstrate purposeful rather than random behaviors and this improves their survival. In this manner effective simulations have evolved along with the organisms that have them.

It's no accident that the functional structure and operational principles of simulations are essentially identical with those used to design autonomous robots. After all both function in the same real world governed by the same laws of nature. And biological organisms function as living purposeful robots in the sense that they autonomously compute their actions. Even details like consciousness and self-awareness are ultimately just a matter of the number and organization of feedback circuits.

Thus the design of the simulation is the basic systems design for any successful autonomous organism and can be used to design artificial intelligence organisms and their interactions with programmed environments in computer simulations. This book deals with the functional structure of the simulation. How it's implemented in biological structures in the brain is a separate issue yet to be fully deciphered and tangential to the subject of this book.

In order for a program to act purposefully it must be able to encode the elements of its purpose and of its environment, and how to effect that purpose within its environment. The simulation is the entire computational control system of an organism that enables this, though its model of reality is of particular relevance here.

The major computational components of a simulation that enable an organism to operate autonomously and purposefully are:

1. **Boot up system**. The basic software encoded and passed in DNA to new organisms. Contains all the code used to construct and maintain the new organism including its body and operational system, which are effectively its hardware and software.
2. **Instinctual imperatives**. Also encoded and passed in DNA these are the fundamental principles that guide the purpose and actions of the organism. The basic instinctual imperative is to generate actions that result in feedback the simulation valuates as positive and avoid feedback it valuates as negative. [The encoding and generational transmission of the necessary operational software of organisms in their DNA is an obvious insight from the computational perspective of Universal Reality. However it appears to have been totally overlooked by evolutionary biologists.]
3. Operationally this translates into a complex hierarchy of instincts, primarily those of survival and procreation. Under survival are the instincts of eating, drinking, and avoidance of predators and personal injury. Specifics variants include suckling and bonding with mothers in baby mammals, quickly learning to walk and run in newborn herbivores, nest building in birds, and seeking the sea in newly hatched sea turtles. Newborn organisms often exhibit specific instincts replaced by others as adults. There is also a vast range of species specific instinctual mating behaviors that produce

feelings valuated as positive in different organisms. These instinctual imperatives define an organism's purpose at all stages of its life. Conflicting instincts may lead to one overriding another when valuated higher as in male-male fights over mating privileges when avoidance of injury overrides the desire to mate. Instinctual imperatives are largely fixed but putting them into action generally involves complex intelligent behaviors.

4. **Data input system**. These are the sensory and perceptual systems of an organism that receive data from external sources and internal feelings from within through the nervous system. The data input system acts as tuned antennae and perceptual filters to extract information of potential value to the organism from the huge mass of raw data in the environment. These systems are especially tuned to detect motion, predators, prey and other food sources, and family or groups members from the background. This involves extensive pattern recognition to organize the visual field into things, processes and dimensional relationships. It includes recognizing the meanings in animal cries and odors, and food tastes, and in humans it also includes the pattern recognition of written words and symbols. This information is then presented to the control system for processing.

5. **Control system**. This is the central operational system that continually receives and processes new input data and acts upon it to update data storage memory and to plan, valuate and generate actions in furtherance of the instinctual imperatives.

6. **Data storage system**. This is the memory of an organism. It consists of short-term memory (data of the conceptual present moment), working memory (data being used in the current computational process) and long-term memory (archive of historical information). Data is generally stored as organized information structures in the form of things, characteristics, relationships, processes, and events. This includes the information of the internal concept of the self within an external world. The underlying organizational principles of this data are stored in the form of the underlying rules by which the world and its inhabitants are perceived to function. These constitute *the logic of things* that are the basic rules by which organisms understand how they and the world operate. It also includes the belief systems and stereotypes organisms see the world in terms of that simplify it and make it easier (though often less accurate) to understand and compute. In general things are stored in terms of their perceptual characteristics and are accessible by individual characteristics. One aspect of an information structure will often retrieve the entire structure. Verbal, or symbolic, memory structures are stored

keyed to the general structures of representational memory.

7. **Learning system**. The learning system continually reorganizes the data input and storage systems and attempts to improve the richness and logical consistency of relationships by extracting, analyzing, and updating their organizational principles. This enables the organism to continue to learn and thereby improve its understanding of the underlying logic of things by which things appear to operate. However there are limits to the degree to which the data storage system can be improved by the learning system, since beliefs and other data structures formed during childhood are often resistant to being changed by contradictory data.

8. **Emotive system**. This system generates excited emotional states of various tonalities in response to circumstances valuated as important. Adaptively this system is designed to energize specific instincts to further appropriate responses. However if negative triggering events can't be rectified this systems may become maladaptive leading to anxiety and depression.

9. **Valuation and meaning system**. Continually compares aspects of stored information against the instinctual imperatives and perceptual feedbacks to valuate and prioritize them in terms of positivity or negativity for the organism. This involves analyzing and assigning meanings to data structures and potential actions.

10. **Planning system**. Imagines and projects future states and different possible actions within them. Plans actions based on comparative meanings and valuations among imagined possible occurrences and actions and their probable effects. This involves intelligent analysis and decision making based on the logic of things.

11. **Action system**. Translates plans into bodily actions from the simplest bodily movement to thought plans involving multiple sequences of actions. Actions in progress are actively refined by perceptual and motor feedback to the planning and control systems.

12. **Focus of attention routine**. This is a specialized subsystem that monitors activities in the whole simulation and informs the organism that it's experiencing those activities. It creates the information that threads of representational information in the simulation are being experienced. The focus of attention routine scans the simulation like an adjustable spotlight with a central focus and an area of peripheral illumination. This routine is the basis of consciousness as explained in the upcoming chapter on *Existence and Consciousness*. It's a largely *linear process*.

13. **Biological processor**. This is the internal biological clock that drives and coordinates all the computational processes of the

organism. It produces the experience of everything taking place within a flow of clock time. This is a *massively parallel processor*. There is too much information being computed within the simulation for it to be otherwise. In turn it's driven by the universal processor that recomputes all the data states of the entire universe simultaneously including the simulation.

All these computational aspects of the simulation are integrated with those that compute all the processes of the somatic body. These are all the actual particle interactions that emergently produce the functions of cells, organs, muscles, and the operations of hormonal, circulatory, digestive, pulmonary, reproductive, and repair systems. Thus the simulation system based in the nerves and brain is overlaid on and becomes an integral part of the computational processes of the total organism. Together they constitute the total program of an organism.

All these systems work together in computational harmony to operate the entire organism. Of course there are innumerable variations across species and individuals from the beginning in single celled organisms up through advanced organisms such as humans and everything in between.

Even single cell organisms are programs and information structures that include some information of themselves within their environment and computational systems that enables them to function better than random chemistry and greatly improves their survival.

This basic systems design including the simulation has evolved over time because it works to improve the survival of living organisms. Thus all living organisms are characterized by having some variety and level of simulation that contains information about themselves within their environments that enables them to compute purposeful actions that improve their chances of survival.

All the systems of the simulation operate in tandem as they continually compute multitudes of simultaneous processes mainly at *the unconscious level*. Only the focus of attention system brings specific operational threads of the simulation into consciousness. It does this by selecting and monitoring specific computational processes in the simulation. This generates the information that process is being experienced, which manifests as the consciousness of that experience.

Everything is the data of what it is. The data of the simulation is the operation of the simulation absent conscious experience of it. In contrast the data of the focus of attention routine is that some aspect of the simulation is being experienced. Thus the reality of this data is the *conscious experience* of that aspect of the simulation.

ILLUSION & REALITY

Though the simulation is essential to our functioning it completely misrepresents the actual nature of reality. The entire world we experience around and within us is a simulation that exists as a running program in our brain. Every one of us experiences reality entirely through its own simulation of reality. Thus the entire physical, dimensional, and biological world we see surrounding us actually exists entirely within our own brains as data within our simulation in a continual process of recomputation.

In general the internal logic of the simulation is sampled from the actual logic of reality, but the appearances of things in the simulation are *qualia* that are painted over the logic to make it appear more meaningful, and easier to valuate, and thus easier to compute our functioning within (Wikipedia, *Qualia*).

Because we experience reality only through our highly illusory simulation of it we have no direct experience of the true nature of reality itself. Thus to understand and experience the true nature of actual reality we can only approach it through our simulation.

Thus to know the true nature of actual reality we must understand how our simulation distorts and conceals it. It turns out it does so in a number of specific ways. By recognizing all the ways our simulation misrepresents reality we can progressively remove these *veils of illusion* one by one until the true nature of reality they conceal is revealed before us in all its glory. Thus to understand reality we first need to understand the illusions our simulation presents to us as reality.

PERSONAL ILLUSIONS

We are all our running programs. Most of the systems and details of our program are more or less fixed products of our biological evolution and individual ancestry encoded in our DNA. But a significant portion of our simulation has been extensively programmed by our parents, peers, media, and other aspects of our cultures since birth.

As a result we inevitably view our world through the lens of this *personal programming*. Our personal programming is specific to us and within our own minds rather than being actual characteristics of objective reality. After all we all see the world differently from our personal perspectives but there is only one common external world we all share. Therefore none of our different personal perspectives can be objectively true.

Our personal programming includes the full spectrum of often misinformed and even delusional belief systems, prejudices, ideologies, ethnic identities, religious beliefs, gender affiliations, and political and interest affiliations. All these become incorporated into people's simulations of reality and projected back onto it as if they were actual attributes of external reality when of course they are not. These heavy personal overlays make it increasingly difficult for people to recognize the true nature of reality within their overly cluttered simulated worlds. And because people think their simulations are actual reality they automatically tend to act in accordance with their personal programming.

Thus our personal programming includes all our personal worldviews, beliefs, neuroses and prejudices. And it includes all the *desires and attachments* that Buddhism rightly identifies as the source of emotional suffering. External reality is essentially neutral with respect to our existence, but it's often thought of as acting judgmentally and emotionally towards us.

Also included in our personal programming is the notion of an objective good and evil, an objective morality. Morality and ethics are social constructs that have evolved because they help stabilize societies. These and other aspects of human culture are adaptive because they tend to damp down the interpersonal conflicts that tend to destabilize social order. Individual ethics and morality are generally learned or reactive variants of the current social norms. The illusion comes when ethics and morality, good and evil, are mistakenly considered objective attributes of reality either religious or otherwise. While such objectification may help stabilize society it's false in itself.

Thus the social norms and belief systems of the society, culture, group, and family an individual grows up in tends to heavily influence personal programming. But in all cases these are attributes of those entities rather than objective attributes of the universe itself no matter how much an individual may believe otherwise.

In essence there are two kinds of people, those who realize they have been programmed and try to transcend their programming, and those who think they are their programming. Unfortunately most people fall into the second category and this is the source of much of the conflict that bedevils the world. Such people have little chance of realizing the true nature of reality even though it lies clear around them.

Thankfully because our personal programming is learned behavior it can be reprogrammed by proper education to reduce dysfunctionality and bring it closer to objective truth. This is the only category of illusion that can be directly reprogrammed. So there is always hope given the proper methods and the willingness to change. There are many methods to help positively reprogram our personal programming but the basic method is to recognize the actual thought chains involved including how they negatively affect us and then break those chains by replacing the weak links with links that connect to more positive chains.

When we realize we are the program code of our running program we find we have considerable ability to reprogram it to what we want it to be and make it happier, healthier, more effective, and successful. We can potentially change our personal programming as much as we want, within the constraints of reality of course. We can reprogram our personal thought processes but not the laws of nature.

The other successful method is simply to ignore dysfunctional personal programming so that it diminishes in importance and begins to fade away. By recognizing the presence of an active chain one can just let it pass on its own without attaching attention or importance to it. As humans we all have desires and attachments, but the trick is not to be attached to our attachments.

SPECIES-SPECIFIC ILLUSIONS

All species view the world through their own sensory and perceptual systems, and conceive of the world in terms of their own conceptual structures and these vary widely among species. Thus every species views the world significantly differently from every other, and every species believes that the real world is actually the way its simulation portrays it.

Obviously this can be true of no species. Our senses are tuned to only a very minute fraction of the information flooding the actual world. We are blind to the rest though some other species have additional or enhanced sensory abilities and our scientific instruments greatly extend our range to sense remotely and view in other wavelengths. Thus the reality of the actual world around us is enormously richer and more complex than our simulation portrays it and it's very different from species to species.

Many species have no color vision at all but instead see much more clearly in low light than we do. The mantis shrimp sees several times as many colors as we do, and eagles see in much higher resolution. And many birds and insects see infrared colors we are blind to, so much so that flowers have evolved patterns that only show up in infrared to attract them.

Likewise elephants and great whales can communicate in low frequency infrasound that our ears can't hear, and numerous species from bats to birds and insects and even dogs hear ultrasound well above the frequency that humans can hear. Thus their simulations of reality include all the information and processes of these additional frequencies.

And of course many species' olfactory systems are much advanced in their ability to distinguish odors over ours. We are near the bottom of the spectrum when it comes to odor detection. And numerous organisms are finely tuned to detecting odors and pheromones of their own species we are blind to.

In addition other species have completely different sensory systems like the lateral lines of fish that detect electrical currents and vibrations or the heat sensing pits of vipers. And there is no doubt at all that the sensation of touch varies enormously from mammals, to snakes, to insects and jellyfish. All of these enhanced sensory capabilities are used to construct an organism's internal image of the world around it.

Thus the simulated worlds each species lives within are all vastly different from each other. And since each individual of every species thinks its own very unique simulation of reality is the way reality actually is, each species experiences reality in a completely different way and thus none experiences reality as it actually is.

Even though we may know better intellectually we can't help but to mistakenly attribute the unique way we experience reality to reality itself. We and all other species have evolved to do this over millions of years. Intellectually we may know better and try to imagine a world of all possible appearances from the view of an omniscient god but even this falls short because actual reality simply has no appearances at all.

THE ILLUSION OF APPEARANCES

Though it's quite obvious that sensory types, ranges, and resolution vary widely among species the whole notion that the universe even has appearances is a complete illusion. External reality itself simply has no colors, odors, sounds, tastes or touches. These are all added by our simulations and painted over the simplified logical structure we extract from reality to make it more interesting and meaningful and easier for us to understand and compute.

We think of the world around us as a colorful world full of sounds, odors, textures and tastes but every one of these is added by our minds to our simulation of reality and doesn't exist in reality itself. Of course photons with various frequencies exist, particle structures transmit sound waves, and molecules carry the information we interpret as odors and tastes but it's our simulation that interprets these as the sensations we experience.

Our simulation creates appearances from our interactions with the raw information of reality, paints these over the information structures, and then projects them back out into a 3-dimensional world it itself constructs. Our simulation then tries to tell us that simulated things out there in its simulated world are real actual things that have colors and textures, and produce sounds and smells when the entire world of appearance is actually produced in our brains, and produced very differently by the brains of other species.

Thus all the appearances of everything in the world without exception are illusions produced by our simulation of the world. Our simulation tells us they are characteristics of external reality but every one of them is produced by and exists only in our own simulation of reality. Actual reality and everything within it simply has no colors, textures, odors, smells or tastes. The world of appearances is a complete illusion.

So when we subtract all appearances from reality we are left only with running programs computing data confirming what multiple other threads of evidence suggest. If actual reality has no colors, smells, tastes, sounds, feelings etc. and these are all in our mind's representation of how we interact with it, then all that is left of actual reality is the information that produces qualia when our own program interacts with it.

Cognitive science is well aware of the fact that all the appearances of the world are internal to individual minds. It even refers to these unique internal appearances as *qualia* (Wikipedia, *Qualia*). Qualia are all the private internal qualities of things, such as colors, feelings, touches, odors and so forth, which exist only in our mind's representations of our interactions with reality rather than in external reality itself. They all exist privately in our individual simulations, and how we actually experience them is ultimately unknowable to others, though we can assume similarities based on similarities of biological structure and our ability to communicate them.

However cognitive scientists never seem able to take the leap in understanding what this implies about the actual nature of reality itself. Typically they go only as far as recognizing that it's impossible to tell for sure that one person's experience of blue is the same as another's which is only the very tip of the iceberg.

This is another of the many ways in which our simulation of reality differs from actual reality. The essential aspect of all of them is that our simulation is not just a model of reality itself, but almost entirely a model of our *interactions* with reality. What we see when we look out into the world around us is not just a representation of the world around us, but everywhere our interactions with the world projected back on to it.

SINGULARITY ILLUSIONS

As individuals we are at only one place and time at once, in only one orientation, scale, spatial motion, and perceived clock time rate at once. But reality itself has no such constraints. This profoundly affects our ability to experience the actual nature of reality.

As individual observers we experience the world entirely from the perspective of our own location and our own particular characteristics. We view it from the singularity of ourselves both as to its dimensionality and its meanings and significance. As a result we naturally conceive of reality as existing in terms of perspectives relative to us but of course this is not the case. This is a subtle but very important distinction.

In reality the observable universe exists from no perspective at all and from all possible perspectives at once. It's perspective impendent but able to accommodate all possible perspectives. What this means is that the observable universe is what is called a *frame independent structure*.

Now the only frame independent structures are logico-mathematical. In fact the great beauty of the equations of relativity is that they are frame independent. This means that everything is expressed in terms of relationships with other things rather than with respect to a single universal reference frame. This allows any particular observer to impose his own reference frame on it, and all frames are equally valid views of it even though they may be incompatible with each other in some respects.

[Relativity asserts the laws of nature of are the same in all frames. However as we have seen in the chapter on *Relativity & Spacetime* this is not entirely true because there must be a preferred computational metric to explain absolute rotation and absolute velocities along world lines. The correct interpretation is that the laws of nature transiently appear to operate the same in all frames from every observer's perspective but their actual persistent effects depend on reference to the underlying computational metric in which they are computed.]

Practically speaking what frame independence means is that the actual reality of the universe is not as it appears from the perspective of any one observer singularity including our own but is such that it makes sense from all and any perspective even though there can be inconsistencies among perspectives. Thus the positions, scales, orientation, motions and perceived clock time rates of objects within the universe are all characteristics seen by individual observers relative to themselves and vary widely among observers.

And in addition the universe *itself* has no intrinsic position, scale, orientation, motion, or perceived clock time rate. The universe has no position, location, or motion because it itself is the ultimate computational source of dimensionality. Thus the universe creates dimensionality among its computational entities but has no dimensionality of its own relative to any external reference frame because none exists.

For example the universe has no scale because the perceived scale of the universe depends on the perceptual scaling of the observer that views it. This will be much different when viewed by an ant or an elephant. Thus the sizes we see things having are simply not actual aspects of objective reality.

And the universe has no orientation since orientation depends mainly on the gravitational orientation of the observer viewing it. The orientation of the universe will be upside down as viewed by observers on opposite sides of the planet though locally both will orient with respect to the surface of the earth. And astronauts aboard the ISS in zero gravity will orient as needed to their workspaces. Thus the universe itself has no intrinsic orientation though we always think of it as having one.

Every organism has its own internal biological clock rate and perceives the clock time rate of events in the world around it relative to that internal rate. Thus fast reacting organisms like flies, chipmunks, and many birds perceive events happening at a slower rate than we do. And elephants and great whales perceive them happening at a faster rate than we do relative to their own internal clock rate. The perceived internal clock rate is roughly related to the size of organisms and the time it takes for nerve impulses and actions to move through the body.

Thus all these dimensional aspects of the universe we take for granted as being intrinsic to the universe are actually observer dependent. Objects in the actual universe simply don't have positions or locations, scales, orientations, or perceptual clock time rates at all except relative to each other or an observer. These are all aspects of individual observer simulations that simply don't exist in actual reality.

Thus as observers we conceive of reality in terms of our own singular perspective. Of course we have some ability to change that perspective at least with respect to position and orientation, and scale through microscopes and telescopes, but we always think of reality in terms of perspectives. But actual reality has no such innate perspective. It's everywhere and nowhere at once, in all orientations and none at once, has every possible scale and none at once, and has every possible perceived clock time rate and none at once.

Another aspect of the singularity illusion is that we tend to think of the world in terms of a few ongoing individual processes at a time since we experience it that way but actual reality always consists of uncountable myriads of ongoing processes. So actual reality is again profoundly different than it appears.

THE ILLUSION OF THINGS

We know from studies of developing minds and the science of robotic intelligence that the identities of individual things are laboriously constructed in the mind from complex repetitive associations of sensory inputs such as colors, textures, forms under rotation and translation, behaviors, functions and other pertinent characteristics. The concept of individual identifiable things develops fairly rapidly in childhood (Piaget, 1956, 1960) but has taken much effort over a number of years to begin to perfect in robotic systems (Wikipedia, *Pattern recognition*). Individual things and events as we perceive them are clearly not necessarily an intrinsic characteristic of actual reality.

Instead the existence of individual things, events and processes is a construct of our simulations of reality based on computational domains as explained in the chapter on *Emergence*. Individual things are emergent entities our perceptual and conceptual systems are tuned to extract from aggregates of computationally associated particles. The concept of a reality composed of completely discrete individual things is largely an fiction of our simulations because at the level of their elementary particles the boundaries of things are in continual interaction and transition and never perfectly distinct.

Nevertheless, at the classical level of multicellular biological organisms, the simulation's representation of a reality consisting largely of individual things works quite well. Biological organisms function quite effectively on the basis of the emergent logic of things that describes the classical world and almost all of science is based on its laws as well. However actual reality is quite different, as it has no such preferred thing-oriented scale, but includes all scales at once. Thus the world of individual things we seem to see around us is simply not a representation of the true nature of reality.

What actually exist are computational domains. Domains are emergent areas of computational density in the universal entanglement network and observers tend to base their concepts of individual things on natural domain boundaries. However domains overlap both hierarchically and interactively so there are no precise actual individual things existent in reality, with the exception of the most elemental. At the emergent level there simply are no exact individual things or programs computing them, there is only the universal program within which domains exist as intrinsically fuzzy overlapping areas of computational density.

Thus a surfer views the ocean in terms of individual waves, a smelt experiences it in terms of tides, and an oceanographer in terms of currents. But these are all domain-based views of a single interconnected ocean, and they all overlap. Leaves, leaf lobes, twigs, branches, trees, tree species and forests are another example of hierarchical overlapping domains that humans selectively identify and compute as individual things on an *ad hoc* basis.

Humans, and no doubt other species, tend to view the world in terms of individual things, properties, events, actions and relationships. These are the basic elements of the logic of things. And they and their logic are also encoded in the elements of grammar with which humans describe their concepts of reality (Chomsky, 1965). This very simplified world is very much easier for humans to compute than the actual world of enormous fluid computational complexity with its continually overlapping domains and programs.

So we humans see the world in terms of individual things and their characteristics and interactions but this is not the actual reality of the world around us. It's another convenient illusion that makes it easier for us to live within an actual continuous reality of programs and data. Our simplified cartoon simulation operates on the basic of the emergent logic of things, but this is far from the computational logic of reality that actually computes it.

This is another example of how the world our simulation constructs is not the true nature of the actual world. The great miracle is the *super-consistency* of the universal program that enables us to function effectively on the basis of the emergent logic of things when discrete individual things don't even actually exist.

So the whole notion of reality consisting of individual things is at least partly an illusion based in the classical scale of humans and other organisms. And there are other critically important ways in which our simulations don't accurately represent external reality.

This is clearly reflected in the nature of the actual moment-by-moment reality of our sensory experiences. What we actually experience are the elemental computational *components* of things rather than the things they are assembled into in our simulations. Only in our simulation do we organize the transitory sensory components of things into discrete things with duration in time.

Things are constructed in our simulations from meaningful associations of sensory components just as they are in robotic systems. The collocated perceptual information of colors, textures, sounds, odors, weights, meanings, and perceived uses all combine to define individual things in our simulations. If all such data elements move together and transform predictably according to the logic of things then our simulations combine them into individual things with those properties. But it's actually the sensory data of the properties themselves that are more fundamental components of reality and even those are emergent properties of particle interactions in aggregate only recognizable by classical level observers with artificially lengthened short term memories able to recognize, compare, and combine them into meaningful things in their simulations.

Even the visual appearances we attribute to things are constructed in our simulation only because the lenses of our eyes focus the blurred mass of background radiation into images. The actual reality of this information carrying radiation isn't focused until the lenses in our eyes focus a small spectrum of it into repeatable patterns the mind is able to pick apart and recombine into individual things. But actual external reality absent our eyes simply doesn't consist of images of things. The true nature of the world around us is a computational space consisting only of the interactions of particle data that has no appearances at all.

THE ILLUSION OF SELF

One of the most significant 'things' constructed by mind, at least for human observers, is the concept of an objective self. Rather than being an actual component of reality our objective self arises in our simulation like all other 'things' do as a fairly consistent fuzzy, flexible and repeatable association of perceptual elements. It's constructed from repeating occurrences of specific categories of associated perceptions that move with our sense of subjective self including our proprioceptive feelings

Associated with our concept of a physical body are models of ourselves in terms of our emotive and thought histories and our memories of actions and how we evaluate ourselves. All this constitutes an internal model of how we conceptualize ourselves as a unique special thing within a world of not-self things.

Only gradually does the division of reality into self and not-self arise in children usually in late infancy (Piaget, 1956, 1960). And only gradually does the realization arise of a self that 'has' perceptions of all the thing constructs that make up both the self and not-self. It's thought that in many other species this process never develops to the extent it does in humans but nevertheless they manage to function quite effectively in their own simulations of reality.

So the notion of self doesn't seem to be a necessary aspect of reality. However it's quite clear that most organisms must categorize their own characteristics separately from those of not-self things on the basis of actions, feelings and meanings. If they didn't distinguish self from not-self aspects of reality in some manner they obviously couldn't survive or function as all their instinctual imperatives are dependent on this distinction.

In the actual reality of the present moment the flow of raw experience occurs antecedent to its discrimination and categorization into individual things and specifically prior to its categorization as parts of self or not-self things. Thus we can reasonably think of all experience without exception as part of a greater or 'true' self that consists of the whole of an observer's experience without exception. This is of course obvious when one considers that all experience, even of seemingly 'not-self' things is actually part of the observer and that the entire experience of the external world actually exists only as experience within the self of the simulation.

From this perspective every experience an observer has is actually part of that observer and what defines it. Whether these experiences are then categorized in the simulation as part of self or not-self they are all part of the total observer experience and thus part of the whole being of the observer prior to the distinction it makes between self and not-self.

Thus every detail of every experience is properly part of the 'self' of the observer because every bit of it without exception occurs within the observer as part of its simulation, and thus the entirety of experience is the true self of the observer.

This is compatible with the view that consciousness itself rather than the fleeting individual contents that appear within it is one's 'true' self but this is just a matter of perspective. From an external perspective consciousness is reality from the point of view of an individual observer, but from the point of view of that consciousness itself there is no individual observer, there is just consciousness and everything that appears within that consciousness is part of its experiential self including the entire external universe as well as the objective self of that observer.

So things are constructed by our simulations out of raw sensory data and encoded and stored as objective concepts and the objective self is one of those things. In our direct experience there is only conscious experience and no objective self and the objective self must be constructed by our minds as a concept we then identify with our subjective self of direct experience (Piaget, 1960).

In particular this includes the concept of self as the physical body. It's likely the strong human sense of an objective self in the form of a physical body only developed to its current level with the advent of mirrors, and later photography and video, which allowed people to see themselves objectively from the outside, a view they rarely had previously. With the current flood of personal images of everyone, especially those deemed most beautiful, has come a much stronger identification of self with the physical body as demonstrated by the modern obsession with personal appearance. This tends to obscure the fact that our true self is the totality and harmony of our directly experienced inner feelings rather than our visual image and objective concept of our self, which are merely mental constructs.

Prior to our current obsession with objective self we thought of ourselves much more as animals do in terms of the subjective self of our direct experiences, perceptions, feelings, actions and thoughts as experienced from the inside. That is of course much closer to our true identity, which is the totality of our direct experience.

So our concept of ourselves as an objective thing with a physical body is as much an illusion as are all the other emergent things in our simulation. These are concepts that are useful in computing our functioning in reality but they are fundamentally misleading illusions that obscure the true deeper nature of reality.

THE ILLUSION OF AN ENGLISH REALITY

We tend to think of reality in terms of the words we use to describe it in our native language. Rather than just seeing it as it is we see it with invisible labels attached to things within it. Thus we imagine the actual structure of reality is largely isomorphic to how we describe it in language but this is far from true. We tend to see the world in terms of the grammatical structure of our language as nouns characterized by adjectives in states and actions as verbs modified by adverbs.

The deep universal grammatical structure of language arose to express the deep emergent logic of things in terms of which humans typically conceptualize reality at the classical scale of their simulations (Wikipedia, *Universal grammar*).

When we understand how language encodes meaning this becomes quite clear. Language uses exact single words to stand for enormously complex data structures that are inherently amorphous and overlapping. Language is largely isomorphic to how we simulate reality as a world of things but bears little resemblance to the actual structure of reality. This is the unrecognized source of many of the difficulties humans have had in understanding reality. Language is designed to describe the simulated world of things rather than the actual reality that underlies it.

Even so language is actually a very poor approximation for representing even the simulated world of things. Most of our information about the structural nature of forms comes from understanding non-verbal forms of meaning. Non-human species clearly have extensive and expert knowledge about their environments including the states, actions and relationships of things in great detail. However since these animals lack the complex symbolic languages of humans how is that knowledge organized and stored since it's clearly not stored verbally?

Humans tend to think that their knowledge of reality is verbal, or at least primarily verbal, but there are many aspects of human knowledge that are similar to that of other species which isn't surprising since human verbal abilities are overlays on the more primitive representational knowledge we share with our animal relatives.

As with animals most human knowledge is stored as organized representations of perceptual memories. Recognition of individual faces is a good example. Humans have the ability to distinguish far more individual faces than they are able to describe verbally and this is true of many other types of knowledge as well. Clearly tagging a face with the name of the person is one type of knowledge but while the name of the person can call up the representation of the face it's clearly not the description of the face, which is stored separately and merely labeled with a name in a type of identity relationship.

Again it's artificial intelligence that sheds light on how knowledge of this sort is extracted from raw sensory input and stored. Facial recognition systems extract and combine measurements of a number of standard features of human faces deemed most useful in distinguishing individual faces to seek the best match from databases of known faces (Wikipedia, *Facial recognition system*). Humans no doubt use a functionally similar method to identify faces though they are still much better at it than computers.

The point is that 'Bill's face' is not stored verbally as those two words but as the association of representations those two words label. The actual stored representation in human memory is an extremely complex set of associated individual data, which may include how Bill's face changes with his various emotions and how it has changed through the time Bill has been known as well.

Thus most of human knowledge consists not of verbal structures but of very complex representational structures built up from organized perceptual complexes. Animal knowledge consists primarily of this though there is certainly a considerably symbolic verbal overlay as well since many animals do associate particular calls and cries with specific representational knowledge.

Many animals express feelings vocally and such vocalizations are essentially the symbolic expressions of those feelings. Such vocalizations also communicate those feelings to other animals quite effectively even to animals of other species. Take the growl of a dog for example or the alarm cries of many species. Feelings arise in response to specific environmental information and can be expressed by vocalizations, which thus are the words, or phrases that symbolically communicate that information along with the feeling it elicited.

The warning cries of birds to the presence of a specific type of predator are effectively the words for the presence of that type of predator in the language of the bird that utters them. And other animals clearly do understand each other's language to a considerable extent, often much more fluently than humans do.

Human language is just a further development. Early humans would simply have vocalized the feelings elicited by social situations such as hunting or other group activities in more and more detail as they performed them with the meaning reinforced by accompanying body language as other animals also do. In this way symbolic language would have gradually developed as a natural outgrowth of animal vocalizations.

The point is that our simulation's model of reality is mainly representational rather than verbal. Though it does, especially in humans, include a much smaller overlay of language structures in the form of the syntactical logic of language, its vast substratum consists of very complex representational structures. Individual words are convenient shorthand labels for often very complex structured sets of representational data.

This is the internal structure of information forms in organisms' simulations of reality, and thus it's this representational form structure that takes us just a little closer to the actual form structure of objective reality. The objective form structure of reality is clearly not verbal. It doesn't consist of individual verbally tagged 'things' that stand in English syntactical relationships to each other.

Nor is reality representational in the form it exists in other species' simulations of reality. These simulations are extractions of emergent structures resulting from their interaction with a reality that consists more of great fluid masses of continuously interacting emergent programs filtered through an organism's interactions with them.

Thus all of the individual things, actions, properties and relationships that make up our verbal model of reality are useful but artificial internal constructs. Though they exist only in our simulations of reality, they are ultimately based on computational domains in the world of forms through the emergent logic of things that represent them in our simulations.

Take the example of a wave in an ocean. The wave is actually part of the continuous form of the ocean and its precise boundaries and duration of existence are to a great extent arbitrary and observer defined, however there clearly are natural domain boundaries upon which the discrimination of a wave from the ocean can be based.

But things are clearly not their language labels. There are no actual 'waves' in the actual ocean, nor are there the discrete things we label as waves. What actually exist are the emergent domains in a continuous substrate from which surfers and others extract their representations of the 'things' they call waves and then label with the word 'wave'.

With technology humans are now able to construct 'things' with sharper than natural boundaries so they more clearly exist as distinguishable individual things with easily identifiable and useful functions. This in fact is an important function of technology, the construction of forms with precise specific boundaries in a world consisting primarily of less precise natural forms.

Thus even though we see the world around us in terms of people, trees, dogs, and stones, there actually are no people, trees, dogs or stones out there in actual reality. These are all just useful labels we attach to our internal representations of the world that don't exist in the world itself.

THE ILLUSION OF A PHYSICAL UNIVERSE

The evidence the universe isn't a physical world but actually consists of running programs computing its data is overwhelming and has been presented in considerable detail throughout this book.

When we look at the things that make up the world around us they appear to be material or physical objects, but every one of them is actually its information only. The apparent physicality of things is an illusion produced by how our minds simulate the world around us. And with practice it's actually quite easy to directly confirm this.

All we need to do is to carefully analyze anything at all into all the information components that make it observable to us such as its color, shape, texture, resistance to motion, odor and auditory characteristics and so forth. Now just mentally remove those information components one by one to see what's left.

We quickly discover that after all the information content of anything at all is removed there is simply nothing left of that thing. So the apparent physicality of things is simply a combination of the various types of information that our mind tells us makes it a physical thing. Thus we can directly confirm the information nature of reality in our own experience.

One could argue that this just applies to the way we experience things and that there are real physical things beyond producing those experiences but there is absolutely no evidence to suggest that and certainly no way to confirm it. There are clearly actual computational processes independent of our experiences but thinking of them as physical is simply a projection of how our mind represents them in our simulation. How else could the information structures of our mind so convincingly encode them as neural data structures if they weren't actually programs and data?

If we take a moment to think about it this is the way we actually experience everything already. The only thing that can be experienced is some kind of information or other. We just have come to tell ourselves that specific kinds of information associations represent what our simulation labels as physical things. But in reality every last thing we can experience is just some form of information because experience consists entirely of information and ultimately the experience of information is all that confirmably exists.

Thus all the apparently physical and material things in the observable universe are actually just the information of what they are. They are the simply the combined associations of all their different information components. Everything that exists in the entire observable universe is entirely the information of what it is and nothing more.

And by extension this is true of the apparent physical nature of the entire observable universe. After all our entire experience of it ultimately exists only as neural information in our brains. Thus the apparent physicality of the entire observable universe is clearly just labels attached to certain types of information associations.

This automatically explains the putative 'unreasonable effectiveness of logic and mathematics in describing reality' (Wigner, 1960). It's completely obvious that the best description of a computational information universe would be a computational information system.

Consider what 'physical' actually means even in physics. It's just a label for something having mass, energy or dimensionality. But as we've seen dimensionality is simply the logico-mathematical consistency among the dimensional relationships generated by particle events. And mass-energy is simply spatial velocity within that logico-mathematical consistency. So even these last bastions of physicality in science dissolve into the information they actually are.

EVERYTHING IS ITS INFORMATION HISTORY

Not only is everything entirely the information of what it is but also the information of what something is consists entirely of its *information history*. Thus everything that exists is its information history and only its information history. The information of anything at all is the current result of its computational history stretching back all the way to the big bang and the complete fine-tuning. Thus the information of anything at all contains snippets of its entire computational information history and that current information history is all that anything is.

The information of a leaf lying on the lawn is not just the information of the leaf lying on the lawn but the complete computational history of all the forces that gave that leaf its individual variation on the form of its species, and all the exact temperatures and breezes that brought it to the exact condition and position on the lawn it now occupies. And it also includes the information of anything and everything that in any way affected that from the effects of the seasonal rotation of the earth, to the DNA particular to its species. All this information is hidden within the current information the leaf actually is.

Thus everything that exists is packed with information about other things. Everything actually is the combined information of all its interactions with the other things that computed the exact information of what it currently is, and everything is the entire computational history of what it is right now.

Thus everything that exists is a *recording* of some of the past information state of the observable universe. Like all recordings each is from the particular perspective of its past computational interactions. But since everything that exists is a recording of a common universal past from different perspectives these recordings taken in combination or all together encode an enormous body of data about the past.

Only because everything contains recorded information about other things and its computational interactions with them is the information of the universe knowable at all. This is the 'Sherlock Holmes Principle' that is the basis of all personal, scientific, and forensic knowledge. By understanding the information of things and the rules by which it's computed we are able to gain knowledge of other things that aren't directly observable. Thanks to the structure of the complete fine-tuning information forms in our observable universe are packed with information about other forms and our universe is richly knowable.

Only through the Sherlock Holmes Principle does the universe become knowable and organisms like us become able to function within it. Only because things are their information and their information is a recording of the redistributed information of other things does meaningful existence become possible.

Every thing that exists is a recording of the information of other things in its interaction history, and the total information content of the universe is continually redistributed through all its information forms as interactions occur and data is continually recomputed and redistributed.

This includes all the data of the history of the observable universe including the original and still current complete fine-tuning that is recorded in the data of all the individual things that exist.

ILLUSIONS OF TIME

In addition to the absence of any innate perceived clock time rate in actual reality and the variations of perceived clock time rates among species covered in the section on *Singularity Illusions* above there are several other very important illusions in the way our simulation represents time to us.

The first is our sense of historical time, which is of course completely lacking in the structure of actual reality, which exists only in the present moment. It's only organisms with memories of past events that have any sense of historical time. While it's true the present moment state of all forms is the computational result of all their past interactions so that they contain and actually are traces or even recordings of the past, there is no actual sense of historical continuity along sequences of events except in organisms with historical memories. Thus the universe can be said to encode its past in all its current data states but only to be aware of that encoding through the organisms it creates with historical memories.

Second the human brain continually constructs its own internal narrative which is continually brought back into alignment with reality by incoming data streams. An important part of that narrative is the simulation continually plays an expected narrative of slightly future events as if they were actually happening (medicalxpress.com/news /2017-05-human-brain-pre-plays-events-fast.html). Thus we perceive an incoming stream of events that are expected to happen as if they are already happening. This gives us an important adaptive edge in preparing to react to events as they occur. In most cases this predictive aspect of the simulation is reasonably accurate on a second by second timeframe but as a result we seem to live slightly in the future as events occur.

Another way our mind misrepresents time is by artificially stretching the perceived present moment of time into a several second duration when the actual present moment in which events are computed is on at least a femtosecond scale. Only this artificially lengthened perceived present moment gives us time to compare and thus make sense of events as they occur. Only this enables us to understand processes as sequences of events. If we couldn't do this the world would appear totally meaningless, as all events would occur without context. Processes like music and language, which depend on running comparisons of occurrences slightly separate in time, would be completely meaningless.

This artificially lengthened present moment constitutes our short-term memory and is absolutely essential for us to make sense of the world. However since this is a product of our individual simulation, it's quite likely its duration and character varies significantly among species. It's completely absent in inanimate programs and we can safely assume it's rudimentary at best in single celled organisms. Though this insight is apparently original with me (Owen, 2009) its effects among species with different short-term memories seems an important area for research.

THE ILLUSION OF DIMENSIONAL SPACE

How what our simulation interprets as a pre-existing dimensional spacetime is actually constructed by quantum events in the form of dimensional relationships among particles has been explained in detail in the chapter on *Quantum Reality*. And this can also be confirmed by direct experience.

When we observe the world around us we never actually see any such empty space. All we actually see is the representation of things our minds construct in our simulation. These representations are of various sizes and orientations with respect to each other, and it's from their relative sizes that they are assigned positions relative to each other in our simulation. This is done according to how their apparent sizes change with motion relative to us. And again this is learned behavior that develops in childhood with the logic of things in the simulation.

It's well known that the world as we perceive it consists of 2-dimensional images on our retinae and that its apparent 3-dimensionality is constructed by our minds comparing these retinal images. This is also true of the dimensional cues from auditory signals. Other species image the world by olfactory signals relative to wind direction, echolocation, heat sensing, and other sensory cues.

So reality as it's actually experienced isn't 3-dimensional at all and its dimensionality is entirely constructed in our simulations from data comparisons. Thus by carefully analyzing the cues, visual and otherwise, that underlie the apparent dimensionality of reality we can cut through them to discover again that actual reality consists only of raw programs and data in the non-dimensional computational space of the quantum vacuum and confirm that the apparent dimensionality of the world our minds project around us exists only in our own brains.

This is quite easy to understand by analogy. The very convincing 3-dimensional worlds of virtual reality actually exist entirely as numbers and logical structures in completely non-dimensional computer memory. It's only when a virtual reality controller projects these dimensional relationships into a dimensional world in a headset that they appear to exist in an encompassing 3-dimensional space. But this encompassing empty space was never even part of the actual data. It only appears to exist because of the projected dimensional relationships among the numeric data of that virtual world.

Thus it's really quite easy to understand how the apparent encompassing dimensional space around us is the exact same kind of illusion arising from our simulation's projection of dimensional relationships among objects. Thus dimensional space is an illusion produced within our simulation to help us make sense of reality.

Though our mental simulation actually exists non-dimensionally within our brains as the universe does in computational space, it seems to exist as a 3-dimensional external world centered on us. This is because our brains project our internal simulations of reality out into the semblance of a dimensional world on the basis of the information of dimensional relationships among the things it contains. But the 3-dimensional space we appear to exist within is actually an interpolation of dimensional relationships computed by quantum events projected into a graphically displayed 3-dimensional world by our minds.

Thus we must recognize that the 3-dimensional world we seem to see around us is an illusion. What we are really looking at when we look out into the world is the information of dimensional relationships of events encoded in the neural circuits of our brains. This information is most certainly not a little dimensional model of reality, but consists only of sets of dimensional relationships among things. The reality around us we appear to live within is actually a virtual reality of our mind's own making.

THE RETINAL SKY

As a result of all the illusions of the simulation, when we look out into the world around us we are actually looking into the internal workings of our own minds. Absolutely everything we see and experience in the world is actually happening within our own brains. The only contribution actual reality makes to this internal simulation is a little logical information structure and even that is a greatly simplified and heavily modified model of the actual information structure of external reality.

Our mind receives colorless data points, arranges them into color-by-number images and then colors and textures them to construct a bright colorful world which it then dimensionally projects out around its construct of us populating it with things it itself created. Our objective self is just another one of the things our simulation creates within its world. Then our mind convinces us the simulation of reality it created is a real actual universe that exists independent of us and our brain when nothing could be further from the truth.

We think reality is the same as our visual and mental representation of it, but actual observer independent reality itself simply has no appearance whatsoever, it's only a computational structure composed entirely of programs computing data. It has no color but only data our simulation interprets as color. Every last bit of the appearance of the world is added by our mind and exists only as qualia in its simulation in our brain. This includes its apparent 3-dimensional structure; its apparent physicality and everything else save some basic logical correspondence sufficient for us to function.

Thus the entire external dimensional world is an illusion in our own brain. It's a little sample of reality's logical structure painted over with appearances and meanings by our mind and projected outward around us as if it were real. The logical structure fairly accurately maps reality's classical level emergent logic of things, but all the appearances it's colored with are added entirely by our minds.

The world we see around us is a moving painting in the gallery of our mind. It's an all inclusive interactive wraparound virtual reality show with us at its center. It exists only in our own brain and so we are actually observing the workings of our own brain much more than the workings of actual reality. To a fair degree the actual logic of the external world is directing the show, but all the costumes, sets, and significance are produced and staged by our simulation.

Thus we look directly and deeply into our own being when we look out into the world. And if we only look deeply enough at what is really going on 'out there' we begin to see the reality within the illusion.

Even basic biology recognizes that the entire visual world around us actually consists of images on our retinae. Thus the entire visual horizon is actually a view of the backs of our own retinae. The sky we see above us is a retinal sky. We live within a retinal horizon and under a retinal sky. We exist within the deepest recesses of our own brain's simulated reality. And even we and our brain are part of that simulation. So ultimately all that exists is the simulation itself independent of any brain or self.

So the analysis of our simulation and how it constructs the world of illusion we experience leads us inevitably to one conclusion, that the world consists only of the abstract information of what it is being continually recomputed by the universal program of reality. So now that we understand that reality is not as it appears to be we are finally free to discover what the true underlying nature of reality actually is.

EXISTENCE & CONSCIOUSNESS

EXISTENCE

To understand the true nature of reality we must understand both its structural aspects and what makes those structures the real actual things of the universe that have existence and being. So far a revolutionary new model of the structural aspects of reality consisting entirely of programs and data at the most fundamental level has been presented.

Now we need to understand how these programs and data gain their existence and reality. The nature of existence itself must be made clear. The implications of this are profound because a proper understanding of existence immediately explains the nature of consciousness and leads directly to a rational scientific definition of realization. This enables us to complete our Theory of Everything and unify all aspects of reality in a manner consistent with both the principles of science and the most fundamental aspects of our personal experience.

There are four fundamental questions of existence. Universal Reality provides convincing answers to the first three of these questions, and a revolutionary new framework for the fourth. And the vast body of established science progressively fills in the details of this framework.

1. Why does something rather than nothing exist?
2. What is the true nature of existence?
3. Why does what exists exist rather than something else?
4. What does actually exist?

EXISTENCE & DATA

1. It should now be quite clear the universe must be a computational system consisting of programs continually recomputing its data for it to work as it does. The precise details of current states are always the result of logically consistent operations on previous states and the only way this can happen is for them to be

computed. And only data can be computed, and only programs can compute data. Therefore there should be no doubt at all that the universe is a computational system as Universal Reality describes it.

2. The code of the elemental program and other aspects of the complete fine-tuning exist as *virtual data* in the quantum vacuum. And the observable universe the elemental program continually recomputes exists as *observable data* in the quantum vacuum. Thus both the complete fine-tuning and the observable universe exist as data of equal reality in the quantum vacuum.

3. Now all forms of data must exist within some kind of medium to be actual. Thus the quantum vacuum must be the medium of reality within which all the data of the universe exists. Thus the quantum vacuum must be the medium of existence within which all the data of the universe exists and thereby gains its own existence.

4. Unlike an ordinary computer program in the medium of computer memory, or ink on the medium of paper, the programs and data of the universe are the real actual things of the universe. They have reality, actuality and being. They have existence. Thus they must exist in the medium of existence.

5. The only way to reasonably explain how the programs and data of the universe gain existence as the real actual things of the universe is if they exist within a universal medium of existence that gives them their individual existence.

6. This means that all the programs and data of the universe are forms *of* existence that exist *in* a medium or substrate *of* existence. Thus everything that exists gains its individual existence from the universal sea of existence within which it exists as a form of existence. This is analogous to an ocean of water in which waves, ripples, currents, and tides are all forms of water that exist within an otherwise formless substrate or medium of water.

7. Thus what science calls the quantum vacuum is its initial discovery and confirmation of a universal sea of existence. The quantum vacuum is the universal medium and substrate of existence itself within which both the virtual and observable data of the universe exists and from which the data of observable particles can emerge and return.

8. In this view only something called existence exists or can exist, and the programs and data of the universe are forms of existence that exist within this universal medium or substrate of existence.

9. Because everything that exists is a form of existence itself, the single fundamental substance of everything that exists is existence itself. Every individual thing is a data or information form of

existence that gains its individual existence within a universal sea of existence.

10. This means that all the data forms of the universe are 'empty' in the sense they have no individual self-substances other than the common medium of existence that gives them all their reality. All the things in the universe are simply forms *of* existence *in* an otherwise formless medium of existence.

11. Thus the universe is a computational system or program running in a universal sea of existence that continually recomputes its entire data state. And it's the entire real actual universe because it consists entirely of data forms in the universal medium of existence.

12. Because the universe consists entirely of existence and forms of existence at its most fundamental level it can't be physical or material in the traditional sense. Its apparent dimensionality and physicality is an interpretation produced within our brain's simulation of reality.

13. The sea of existence is complete in itself and there is no outside nor before nor after. It is everything and contains everything and nothing else exists or has ever existed or can exist. The universal medium of existence is the true fundamental nature of the universe itself. The sea of existence is the complete universe and all the data forms of existence that arise within it are all the real and actual things of the world that data is the data of.

ATTRIBUTES OF EXISTENCE

1. To account for the reality of the universe existence must have several intrinsic attributes: *presence, happening, absoluteness, xperience* and *immanence*. Existence is also *self-necessitating*, and must be *logically consistent and complete*.

2. To exist existence must have presence. The presence of existence manifests as a *universal present moment* common to the entire universe. Thus existence exists only in the current present moment it manifests by its own presence. The entire universe consists only of its current present moment data state. Neither the past nor future has any actual existence. The presence of existence is the present moment of our experience, one of the most fundamental aspects of our existence.

3. *Happening* is the fixed universal flow of present moment time. Happening is the cycling of the universal processor that

continually drives the recomputation of the data state of the entire universe. It computes all local clock times based on the spatial velocity present. It's the computational source of clock time and all change.

4. The current data state of the universe is *absolute*. It's absolutely real, and absolutely exactly what it is. Once it exists there is absolutely no possibility of it being different in any minute detail whatsoever. Thus there is absolutely no possibility whatsoever that the entire current data state and historical logical structure of the universe stretching back to the big bang could have been any different in the most minute detail. Thus for example there is no possibility whatsoever of alternate fine-tunings. The current absolute reality of the present conclusively falsifies even the possibility of alternate pasts or fine tunings. This answers the third fundamental question of why what exists does exist.

5. All the programs of the universe can be said to experience, in a generic sense, the data forms they computationally interact with in the resulting changes to their own data forms. Experience in this generic sense can be called *xperience*. Xperience is the universal mechanism fundamental to all observation, knowledge, and science and is what we call *experience* when it occurs consciously in our own minds.

6. *Immanence*. The data of the observable universe is absolutely real, actual and present. *Existence actively self-manifests itself* through all its data forms which is what gives them their reality, presence, happening and absoluteness. What we call *consciousness* is the immanence of the data of our experiences. Consciousness is our direct experience of the actively self-manifesting immanence of data forms of experiences happening in our simulation.

7. Existence is *self-necessitating* because non-existence cannot exist therefore existence must exist. This is the single self-necessitating fundamental axiom upon which the entire logical structure of reality stands. Thus existence must have always existed and there is no need for a creator or creation event. There was never a nothingness out of which something appeared. (The big bang was an *actualization event* of observable elementary particle data within existence.) This is the answer to the first fundamental question of existence of why something rather than nothing exists.

8. Because it's computational the data structure of existence must be entirely logically consistent and logically complete. Otherwise a computational universe would tear itself apart at the inconsistencies and pause at the incompletenesses and could not exist. The program of the universe would either crash or hang.

All these points taken together provide the most reasonable and consistent answer to the second fundamental question of what the true nature of existence is.

OBSERVERS AND XPERIENCE

The concept of an observer is implicit in Universal Reality's model of reality. All emergent processes can be said to *xperience* each other in the changes to their own data produced by their interactions. Thus observation in the generic sense of changes to a process's data is an intrinsic aspect of happening and occurs with all the computational interactions in the universe.

When xperience occurs in living organisms' simulations of reality and rises to consciousness it's called experience. Thus *experience* is simply a special case of the universal computational mechanism of *xperience*.

In a sense the universe can be thought of as continually xperiencing itself into existence through its continual computational interactions because existence is ultimately confirmed only through experience. Thus all the programs of the universe can be seen as generic observers continually xperiencing each other and themselves in the computational changes their interactions produce to their data. From this perspective the universe continually xperiences itself through all its interacting processes through all their observations of each other.

Thus the very existence of observers is implicit and central to Universal Reality. In this manner Universal Reality seamlessly integrates the role of observers into its fundamental computational fabric.

1. Everything that exists is the complete data of what it is filled with the immanence of existence that makes it the real thing it's the complete data of and that is all anything is.
2. Computational interactions of things are defined as processes that make changes to the data the things actually are. These data changes may be structural or dimensional (for example positions and velocities) or both.
3. In a generic sense all things effectively experience their interactions with other things in the changes made to their own data. Such generic *experiences* can be called *xperiences*.

4. In this sense all things act as generic observers in their interactions with other things. Thus every computational interaction is a mutual generic observation in which each interacting thing xperiences the other.
5. Thus there is a meaningful sense in which even the programs of inanimate processes can be considered generic observers as their interactions with other programs alter their own data. Thus all interactions can be thought of as generic observations or xperiences.
6. Because the observable universe consists of things or programs in continual interaction we can view this from the perspective that the entire universe consists of xperience only.
7. Because existence is only confirmed by observation we can say the observable universe continually xperiences itself into existence. We can say the observable universe consists entirely of generic observers that continually xperience it into existence by observing it in their interactions.
8. Almost all xperience occurs as changes to the data of inanimate processes. Inanimate programs xperience their existences but have no way to know, record, or report their xperiences because they occur without the meaningful context of an internal simulation model of themselves within their environments.
9. Thus inanimate processes are unaware they are having xperiences because they have no experiences that they are having xperiences. Thus almost all xperience occurs without awareness or consciousness. Thus *xperiences* are effectively *experiences* that occur unconsciously never rising to the level of consciousness.
10. Thus xperience is a universal process that underlies all the computational interactions of the observable universe. However when it occurs in specialized forms in the simulations of living organisms it becomes what we call experience.
11. *Experience* is a subset or special case of *xperience* that occurs in an organism's simulation models of reality. Other than that it's the exact same fundamental process of a change to the data of an organism resulting from its interaction with another process.
12. When the universal phenomenon of *xperience* occurs in the simulations of living organisms it takes the form of *experience*. The difference is that the experiences of living organisms occur in the context of the organism's simulation model of its environment, which gives it the context to be meaningfully experienced and recorded.
13. Thus all the processes of the universe can be considered to consist entirely of *xperience* but only those of the simulations of living organisms are *experience*. So the self-awareness of living

organisms consists of the same xperiences as those of inanimate processes but they occur in the context of an internal information model of reality whereas those of inanimate processes occur without any context.

14. Specifically an experience is the changes to the data representation of some thing or event in an organism's simulation due to its sensory interaction with the program or data of that thing. So the fundamental process involved is exactly the same as that of *xperience*.

15. Because changes to the data representing things in the simulation occur in the context of a model of reality such changes have meaning and are knowable. Thus they can potentially rise to the level of conscious experience if spotlighted by the focus of attention routine.

16. Thus existence's fundamental attribute of xperience enables us to understand the experiences and self-awareness of living organisms in the broader context of the fundamental processes of existence. And to finally explain the true nature of consciousness and how such experiences become conscious experiences we need only to understand the nature of *immanence*, another fundamental attribute of existence.

CONSCIOUSNESS EXPLAINED

The problem of what consciousness actually is has remained an unsolved mystery throughout history and neither philosophy nor modern science has had anything meaningful to say on the subject. The basic problem is that consciousness is clearly not anything physical so as long as the universe was considered a physical structure there could be no solution.

However Universal Reality clearly demonstrates that reality is not at all physical in the traditional sense. Instead it's a computational structure consisting entirely of data, but data given the reality of existence by existing in an underlying medium of existence.

This means that everything that exists is the real actual thing it's the data of because its presence in the medium of existence gives its data the being and reality of whatever it's the data of.

Existence is not passively sitting there waiting for its realness to be manifested in the sense a physical world was assumed to do. Instead it actively self-manifests its presence and reality through each of the data forms that exists within it.

Thus every program and data form that exists actively manifests its existence through its form. This is what we mean when we say something is real and has being or existence. Whatever anything is the data form or program of that data form or program actively manifests the existence of what it's the form or program of. The data form becomes the real actual thing it's the data form of because it exists in the medium or substrate of existence.

This active self-manifestation of existence in all programs and data forms is called *immanence*. Thus the immanence of existence shines with reality and being within all things. The immanence of existence is the real actual presence of things within reality. The immanence of their existence is what makes things real actual existent things.

Immanence is not a new concept. Historically it refers to the hidden presence of the divine in things. Universal Reality uses it in the exact same way but without any religious connotation. Immanence is simply the hidden presence of existence within all things that exist. But of course immanence is not actually hidden because it's what makes all things actually real and present in the universe. Without the immanence of existence within things they wouldn't even be there.

So existence is what makes all the data forms that exist within existence into the real actual things of the observable universe. Everything is actually the complete data form or program of what it is and immanence is the realness manifesting within all data forms that makes them the real actual things they are the data of.

Existence is what makes data forms the real actual things they are the data of, and the immanence of existence is the actively observable self-manifestation of the presence and realness of existence in all things.

Thus consciousness is simply the immanence of the data that something is being experienced by the focus of attention routine in the simulation of a living organism. The immanence of the data of an experience makes that experience a conscious experience. *Consciousness is simply the immanence of existence shining within the data forms of experiences in the simulation.* This internal shining isn't visible to the eyes but is visible to mind as consciousness.

This is the only possible reasonable explanation of consciousness. Consciousness couldn't possibly even exist if the universe was entirely physical because consciousness clearly isn't physical. And the usual non-physical attempts to explain consciousness always seem to be antiscientific and rife with New Age superstition.

However our model is perfectly consistent with established science at the computational level and just uses the self evident and very obvious fact of existence and its essential attributes to explain both the consciousness and reality of things as aspects of the same universal attribute of immanence that manifests data forms as the real actual things they are the data of. Consciousness is just the immanent reality of the data that other data is being experienced.

Thus everything 'shines' with the immanence of its existence. This shining is what we experience as the realness and actuality of things. *The actuality and realness of the data that representations of things in the simulation are being experienced manifests as our consciousness of them.*

It's all very simple, reasonable, and elegant and convincingly solves the problem of what the true nature of consciousness is. Consciousness is simply the self-manifesting reality of our focus of attention routine generating the data that other representational data is being experienced. *The immanence of experience is the consciousness of experience.*

1. Our simulation continually updates its internal data representations of things and processes through sensory data input and reorganization. This data in the simulation is part of the entire data of the whole organism. This continual updating of the representational data of self and environment within the simulation is a massively parallel process that occurs almost entirely at an unconscious level.
2. This representational model of self within environment is what our simulation tells us the real actual world centered on us. The reality of the data of the simulation is the actual simulation model in our brains.
3. There is an additional *focus of attention routine* that scans the simulation selecting individual things and connections of particular interest to the organism.
4. Thus the data computed by the operation of the focus of attention routine is the data that some of the representational data is being scanned, in other words that that representational data is being *experienced.* We then think we are experiencing something in the

external world because our focus of attention routine is experiencing our representation of it in our simulation.

5. The reality of the representational data is simply that it's data representing some thing or process in our simulation. Its immanence of existence makes it the real representational data in our simulation in our brain. It's really there in our brains.

6. The reality of the focus of attention data is that some of that representational data is being experienced. The immanence of the existence of that data makes that experience the conscious experience it's the data of.

7. Thus consciousness is simply the actively self-manifesting immanence of existence of the data that some representational data in the simulation is being experienced by the focus of attention routine. It's our consciousness of an experience of some representational data in our simulation.

8. Our simulation then tells us we are actually experiencing the external thing that representational data represents but we are actually only experiencing its representation in our simulation. However since we are experiencing it consciously we mistake it for being conscious of things in the external world.

9. Thus consciousness is the immanence of existence of the specialized data that representations of things in our mind are being experienced by our focus of attention routine and that's all there is to it. Consciousness is really quite simple once the proper foundation is laid.

In ancient Greece some philosophers mistakenly thought that seeing involved the eyes shining light on things. Today almost everyone still thinks that consciousness somehow involves the brain generating consciousness and shining it onto external things to make them conscious but it's time to consign this theory to the scrapheap of history along with the old extramission theory of vision it's based on (Cornford, 1997).

The correct picture is that our simulation constructs internal data representations of external things. And it also constructs data that some of these representations are being actively experienced by the focus of attention routine. But so far this is just data. What gives all that data its respective reality is the immanence of existence within all these data forms that makes them all the real things they are the data of. Consciousness is simply what the reality of the data that representations in the simulation are being experienced is. Thus consciousness is not something generated by human brains but the fundamental nature of reality manifesting within data structures produced by our brains.

Immanence manifests in all the data forms of everything in the observable universe in the exact same way but in all cases exactly in the form of that data. Thus the data it's the data of becomes the real actual thing it's the data of. The data that a representation is being experienced manifests as the consciousness of the experience of that representation.

So our consciousness is precisely our direct experience of the immanence of existence in whatever we are conscious of. It's the direct experience of the self-manifesting reality of things, of the data of all things that gives them the reality of the things they are the data of.

Thus the essential active ingredient of consciousness exists in everything in the universe. But for immanence to manifest as consciousness it must occur in the form of experience in the mind of some organism. The specific forms of the contents of consciousness depend on the perceptual and cognitive structures of the biological entity but the fact of consciousness itself, that those contents are conscious, is due to their immanence.

The data of the universe is not just data like words on a page or bytes in a computer memory because it exists in the underlying medium of existence science calls the quantum vacuum. Thus the existence of all the data of the universe actively self-manifests it as the real things of the world it's the data of. We can call this active self-manifestation immanence. In this sense all the things of the world glow with the actively self-manifesting immanence of their existence, and what we experience as consciousness is only one example of this universal phenomenon that occurs only in specialized routines in our simulation.

We must also clearly distinguish between the *contents of consciousness* and *consciousness itself.* Consciousness like existence is a realm within which contents arise. So consciousness itself within which contents consciously pass through our mind is our direct experience of the universal formless substrate of existence or at least as close as we can come to experiencing it in human form.

We can better experience this through the simple mental exercise of meditation in which the passing contents of consciousness are allowed to pass without attention so that they gradually subside and fade away. As the contents of consciousness fade away all that's left is the bright clear field of consciousness itself. This is the direct experience of the immanence of existence and this is the key to realization.

REALIZATION

APPROACH

Universal Reality defines reality as the totality of everything that exists, and clearly demonstrates that the true nature of reality is not as it appears. The appearances of both ourselves, and the world we seem to exist within, are adaptive but illusory simulations created by our minds in the form of the familiar physical world of our experience. But this illusory world is nothing at all like the true reality of running programs in computational space other than it shares sufficient logical correspondence to allow us to function in the actual world.

Nevertheless our only possible knowledge and experience of reality is through our mind's simulation of it. Thus to discover and directly experience the true nature of reality we have carefully examined the illusions of our simulation to reveal the fundamental truths they conceal. We are now in a position to directly experience reality as it actually is insofar as that's possible in human form.

Realization can be defined rationally and scientifically as understanding and directly experiencing the true nature of reality that lies hidden within our simulation of it. In Universal Reality there is nothing more to realization than that, and there are none of the usual metaphysical or religious connotations implied. However understanding and directly experiencing the true nature of reality is certainly the most awe inspiring and transformative experience that can be imagined.

The fundamental elements of realization are realizing the illusory nature of the world of experience; realizing the actual computational and information nature of all things; and realizing the immanence of existence in all things, including ourselves. Realizing the world as its running programs, and experiencing the immanent existence of these programs, including the immanence within our own program, is the key to realization.

In addition there are the continual possible realizations of the deeper natures of all the individual processes of the universe including the deeper aspects of the events of our daily lives. The fundamental aspects of realization are explored in this chapter but for the seeker realization is an unending process of better understanding the inner workings of reality in all things including the seemingly most mundane.

The foundation of realization must lie in the best Theory of Everything possible because only a deep, accurate, and comprehensive theory of reality and how it works can reveal the true path to its realization. Universal Reality provides this foundation based solidly in reason and science. Its development over the centuries has involved solving a long series of basic questions including *the relativity and quantum kōans*. This entire book has been devoted to carefully constructing the necessary foundation on which our personal realization can be solidly anchored.

Universal Reality also sheds light on some core concepts from Western and Oriental philosophy and gives them a consistent modern scientific interpretation. This enables some important personal approaches to realization from which basic ethical principles can be derived.

To summarize, realization can be scientifically defined as understanding and directly experiencing the true nature of reality. This includes the following:

1. The realization that our consciousness in the present moment is the direct experience of the fundamental processes of the universe happening within us.
2. The realization of the true nature of our experience of the flow of time and happening.
3. The realization of our existence in a non-dimensional computational space.
4. The realization of the information nature of all things, that all the things of the world are information structures continually being recomputed by the running programs of reality.
5. The realization that the information structures of all things are computational recordings of their information histories dating back to the big bang and the complete fine-tuning.
6. The realization that existence itself is the common fundamental 'substance' of all the forms of existence. That all the forms of existence are 'empty' of anything other than existence itself.
7. The realization and direct experience of the *immanence of existence* as *consciousness*. The realization of *consciousness* as the presence of immanence in all the contents of consciousness and in the original formless nature of consciousness and existence itself.

186

8. The realization of the *primacy of experience*, that ultimately all that observably exists is conscious experience in whatever forms it occurs.
9. The realization of *true self* as the conscious experience of immanence prior to any thought forms.
10. The realization of *chi and energy body* as the fundamental structural organization of direct experience.
11. The optional realization of the *personal myths of God and Buddha Nature*.
12. The *ultimate realization* that the illusion of our simulation recognized as illusion is our only actual direct experience of reality. That we are all already directly experiencing the true nature of reality in every experience and we are all already enlightened.

THE FUNDAMENTAL REALIZATION

The fundamental realization of Universal Reality is that the central experience of our existence, our living consciousness in a present moment through which clock time flows, is in fact our direct experience of the most fundamental process of the universe occurring right here within our own being.

Our consciousness in the present moment is our direct experience of the happening of the universe computing its continual evolution within the presence of existence. We directly experience this because it's continually occurring within us just as it does within everything in the universe. It's our direct experience of the running program of the universe continually recomputing the entire data state of the universe including the data of ourselves.

This is the fundamental computational process that makes us alive and conscious within the universal presence of reality. It's our direct experience of the universal processor continually recomputing our existence into the current present moment. This is the fundamental process of the universe and we are right there in the middle of it experiencing it in every second of our existence in our own being.

The entire universe can be considered a *living system* in the sense that it continually happens on its own with no external cause or motivating force. Thus everything in the universe shares the life of the

187

universe. All the individual programs that constitute the processes and things of the universe are also alive in the sense they are all parts of the living system of the universe itself.

The universe isn't alive in the sense of a living organism but the life we feel within us is the universe living within our form. We glow with the life of the universe continually happening within us. The universe lives within us, and we live as part of a living universe. The universe lives us from within by continually recomputing our data in the current universal present moment of existence. Our life in this very moment is our experience of the universal processor computing the universe within us.

Realizing this is the fundamental realization of our existence. The fundamental process of the universe is not something just happening far out in the depths of interstellar space. It's happening inside every one of us all the time, and all that needs to be done is to realize this and experience it for what it actually is.

THE REALIZATION OF TIME

Our fundamental experience of the flow of time and happening is us continually zipping through time at the speed of light. As explained in the chapter on *Time* everything continually moves forward in clock time at the speed of light on its own clock. Even as we read this line we are continually hurtling through time at the speed of light. This is the realization of the velocity of time.

We also directly realize that the universe exists only in a single current universal present moment because we directly experience it as such and have proved it in the chapter on *Time*. Our entire lives consist of present moment experiences that change as the current present moment is computed from the previous present moment. All our memories of the past and the other computational results of the past exist entirely in the current present moment. And we never experience the future because it hasn't been computed. This is the realization that only the present moment of time actually exists.

Thus we immediately realize the impossibility of time travel in the sense of traveling out of the present moment. The current present moment is all that exists and where everything exists and happens.

However we can directly experience the fact that we live within a 4-dimensional hypersphere simple because we actually see it from the inside. We do see down the radial 4th dimension of past clock time as distance in every direction from every point in the 3-dimensional spatial surface of the universe. We see down the past dimension of time because of the finite speed of light but we aren't observing an actually existent past, but the light from the past in the present moment. This is the realization that we live within a hypersphere universe.

We can also directly realize that clock times pass at different rates through the present moment depending on the amount of spatial velocity present. We directly observe this in the lengthened half-lives of decaying particles moving at relativistic rates and the speed of clocks on earth relative to those traveling in space. Even magnetism is our direct experience of the relativistic effects of moving electric charges. This is the realization of the interdependency of clock time rates and spatial velocity as expressed by the STc Principle.

Finally there is the illusion of the duration of the present moment itself. The present moment of our experience seems to have a sliding duration of several seconds so that our minds have long enough to compare and make sense of things. But the actual physical duration of the present moment in which the programs of the universe recompute their data is far below the resolution of human experience and even far below the time scale of elementary particle interactions.

It's only because our short term memory holds a simulated present moment open long enough for our mind to compare things that anything makes any sense at all. If our short-term memory didn't work this way we would not even be aware of changes as they occurred since that depends on a mental comparison of before and after states in an artificially extended present moment of consciousness. Without this illusion of present moment duration we would still experience reality as completely real but completely lacking context and completely meaningless.

This can be experienced directly to some extent. If we rest with eyes closed and listen to calm music or even a single tone and progressively direct our attention closer and closer to the exact instant that it appears into and out of existence we finally cut through the illusion and experience a state of instantaneity of time, a vanishingly short duration present moment and we suddenly realize the true nature of the present moment of time. It's a vanishingly short instant, and within that nearly non-existent moment lies the entire existence of the universe and us as well. These are the essential aspects of the realization of time.

THE REALIZATION OF FORMS

This is the realization that all the apparently physical things of the universe are actually their information forms and that all forms are empty of any self-substance other than their common existence. This includes the forms of all individual things and the apparent form of a dimensional spacetime within which they reside. And it includes the complete forms of ourselves as well. Every form is the result of a running program that continually recomputes it in the current present moment.

The immanent information nature of things can be directly experienced and it's fairly easy once one gets the hang of it. This is a straightforward procedure that can be applied to anything including ourselves. We need only mentally deconstruct anything at all into every last aspect of what makes it appear to be physical.

It quickly becomes clear that what makes things appear to be physical objects is simply collocated associations of various types of information such as color, shape, hardness, weight, texture, context, and often use and meaning. And if we mentally subtract that information piece by piece we quickly find there is nothing at all left of anything because only information in one form or another is even observable. When all the information of anything at all is removed from it the only thing left is the immanent emptiness of its existence that made that information seem like a real and actual thing in the first place. Thus all things reduce to the immanent information of what they are and nothing more. There is no inherent physicality left when information is removed

Consider the stone by the side of the road. We can easily mentally deconstruct it into what makes it appear to be a physical object. Its visual color and texture are information encoded in our brains about how our eyes and visual systems perceive it. Its hardness and texture are the information of how our muscles and fingers interact with it. Its odor, if any, is the information of how our olfactory system interprets it, and the sound when we strike it is the information of how our auditory system perceives the resulting sound waves entering our ears. Discard all this information and there is nothing actually left of the stone. Thus the stone is the set of all its information and that's all it is.

Thus what we call physical things are simply associated sets of specific kinds of information that our simulation labels as physical objects. And these labels are just more information. So it's quite clear, and easy to realize with a little practice, that all the stones and other inanimate objects of our experience are actually collocated associations of specific types of information that our simulation tells us are physical objects. Their apparent physicality is a label added to the information of a thing so our simulation can make better sense of our environment by categorizing information in useful ways. But all such categories are more information on top of information and finally everything in our simulation consists only of its data.

This is the realization of the true nature of the stone as we experience it in the neural data of our brains, and there is every reason to think that the true nature of the stone in the external world also consists entirely of its information just as our extremely convincing mental representation does. What happens is that our program, consisting only of information itself, interacts computationally with the program of the stone, also consisting entirely of information, to generate the information of our interaction with it, and that information is then encoded in our simulation and interpreted as the physical stone of our experience.

It can still be argued the stone is a physical object at the particle level but Universal Reality demonstrates that even elementary particles and particle components are best understood as data structures. Science has progressively reduced the apparent physicality of the entire universe to just elementary particles and spacetime but even physics agrees that particles certainly aren't physical in the usual sense of the word but are composed almost entirely of empty space and force fields and Universal Reality demonstrates that spacetime is actually the internal logico-mathematical consistency of dimensional relationships among particles. So Universal Reality just takes the reduction of the apparent physicality of the universe to its logical conclusion as consisting entirely of data.

And there is an even deeper realization of the world of forms. This is the realization that the forms of all things are recordings of their past computational interactions back to the big bang. Thus all the information of forms is the redistributed information of other forms and this Sherlock Holmes Principle is the source of all our knowledge of the universe.

Thus the true nature of that leaf on the lawn is the complete information of what it is. But that information is much richer that its immediate appearance, because its exact location on the lawn is the result of an enormously complex interplay of interacting programs that computed it into reality. The exact size and aerodynamic shape of the leaf in combination with the exact breezes that brought it to this precise location, and the exact information of the chemistry that loosed it from its twig at the exact moment those breezes were blowing all interacted computationally to bring it to the exact position it occupies now.

And the moment of separation and its shape and weight are the current computational results of millions of years of evolution of its species that in turn is the computational result of the uncountable program interactions of its evolutionary selection. And the DNA form of that leaf responsible for the general plan of how the tree it came from grew and developed and produced it on the twig it fell from are also essential components of the computations which resulted in the leaf as it lies at this moment on the lawn.

And further back the acorn that fell in the exact spot its parent tree sprouted and grew, and the lineage of all the acorns and trees back through the entire history of the species and before back through the entire history of its plant ancestors, all must have been computed exactly with not the slightest deviation for this single leaf to now lie on this lawn in this exact place and position at this very moment. And all this information is hidden in the information the leaf is as it lies here on the lawn right now. Experiencing this in the forms of all things is the full realization of forms.

Everything in the universe can be considered the running program that continually recomputes its information. We can directly realize this programmatic nature of things by simply analyzing things into their data and watching that data computationally evolve in interaction with the other programs of its environment.

Consider a housefly. The fly is clearly a very active program that generates continual changes in its information. It's a little biological robot with a robust computational system capable of highly intelligent (relative to random behavior) decision-making in accordance with its instinctual imperatives. It samples relevant information from the information of its environment, and computes effective actions to feed, avoid damage, and reproduce. These systems are supported by an enormously complex integrated hierarchical program down through the subprograms of every cell in its body to the particle interactions that ultimately compute them.

It's this complete program of the fly that actually is the fly. Like the stone, its apparently physicality reduces to the computational interaction of our program with its program. We can clearly experience the fly as an intelligent running program that generates the information our program interacts with to generate the information of our experience of the fly produced by our simulation.

Thus the fly, the stone and all the individual things of the world can actually be experienced as the running programs they are as experienced by our own running program computationally interacting with them. To make the data of reality easier to compute our simulation represents it as the familiar physical world populated by inanimate objects and living beings all neatly organized into categories meaningful to our functioning but this is an illusion.

When we actually look at the world with opened eyes we see that every bit of it consists only of the information of what it is. Nothing changes; this is simply a new way to look at what has always been there and what we've always been looking at. The information of things is all we ever experience of them, and information is all that can be experienced, and when this is understood and directly realized the true nature of the world we live in is revealed. Nothing can be experienced other than information. Nothing other than information and its immanence can possibly be experienced, thus all we experience ultimately consists only of information given reality by the immanence of its existence. And because the external world can be so convincingly represented only as information in our simulation of it we can assume with overwhelming certainty it also consists entirely of information.

The realization of forms changes nothing about the world. The world remains exactly as it always was, but now its true nature as the information of everything that it is, and the information of the running programs that are continually computing that information is realized. Now as we look out into the world and into ourselves it becomes clear to the realized mind that all is information only. We actually see the world as the information it is, being continually computed by all the programs of the observable universe including our own. Everything that exists is the running program of itself continually recomputing its information and these are all the real things of the world because of the immanence of the existence in which they all exist and happen. Profound implications follow naturally as we now see the world of forms through opened eyes.

This is very similar to the ancient Oriental teaching that the essence of all forms was śūnyatā or emptiness (mu). These teachings agree that all things are actually empty information forms filled only with formless existence itself and that directly experiencing this is realization.

REALIZATION OF IMMANENCE & CONSCIOUSNESS

All religions have their mystical traditions and in modern times many spiritual beliefs have developed outside of traditional religious contexts. Our new theory of reality leads seamlessly to a scientific theory of realization and a logical explanation of these spiritual and mystical traditions.

The essence of realization and all mystical experiences is simply the experience of immanence. While immanence exists in all things its association with religious symbols in the context of a belief system most often manifests as the presence of a god or supernatural beings even when these are mental constructs that have little correspondence with objective reality. This recognition of the immanence of existence in religious forms is the source of the traditional meaning of immanence as the presence of the divine.

The philosophers of the ancient Indian and Buddhist traditions also discovered the immanence of information forms through a careful analysis of consciousness from within (Tsunemitsu, 1962). The notion of the emptiness of all forms revealing the underlying presence of Śūnyatā or nothingness (mu) is clearly describing the same thing in a prescientific context (Wikipedia, *Śūnyatā*) (Suzuki, 1956).

And certainly the Indian and Buddhist concept of enlightenment is describing a state where the immanence of existence is directly experienced both in forms and in the underlying formlessness they arise within and as the essence of our own being and consciousness (Wu, 2005).

The use of psychedelic drugs can also enhance the recognition and experience of immanence. In particular LSD, mescaline and psilocybin have this effect where the common things of the world take on an enhanced reality, which is essentially the recognition of the immanence they always have but which is rarely recognized or appreciated in daily life.

Transcending its usual religious connotations Universal Reality recognizes immanence as the actively self-manifesting presence of existence in all the forms of existence. It's their existence that gives all the information forms that exist within existence their reality, actuality and being, it's what makes all the information forms of the universe the real actual things they are the information of.

And as explained in the chapter on *Existence & Consciousness*, consciousness is simply the immanence of existence in the information that representations of things are being experienced by our focus of attention routine in our simulation. Consciousness is the immanence of experience.

Thus the realization of immanence lies in consciousness. By simply realizing that our consciousness of things is our direct experience of the immanence of their existence we immediately recognize the presence of existence that gives all things their reality and being.

All the forms and programs of the universe actively self-manifest the existence they are forms of. This is true of our own program and form as well. We continually glow with the immanence of our existence and this is our fundamental nature as it is of all forms. And this can be realized simply by turning our conscious attention to it and feeling it inside us in all the forms of our consciousness. Our own immanence is always there just waiting to be experienced. It's what the experience of the life force within us actually is.

The living presence of existence continuously glows and flows with the immanence of its being within all things giving them their actual presence, being, and happening. We too exist within this universal sea of existence, which gives us our life, our presence in reality, and all the wonderful manifestations of the running program we are, and which we directly experience as our true self if we only stop and realize it.

So the direct experience of this living immanence of existence in all things is central to realization. When immanence is truly realized it's an amazing transformative experience. The world we exist within remains exactly the same as it was before but the consciousness we see it with has forever changed. We become our program running in an immanent sea of existence, and we experience the living existence of the universe glowing and flowing and within us giving life, reality and being to the information of ourselves and manifesting as our consciousness of the profound and glorious universe we live within.

The realization of immanence tends to arise naturally with the realization of things as their information. When things are fully recognized as their information forms then the immanence of their forms naturally shines forth as our consciousness of them.

The immanence of the existence of all things now begins to become clear. We realize that all the information of the world exists in the originally formless sea of existence itself within which it arose and became real, actual and present. That original real, present, and actual absoluteness of formless existence is always there within all information forms including those of ourselves. It's the formless sea of existence in which all things exist and we directly experience it in all the things of the world as our consciousness of those things. This is the fundamental experience of reality and this is its realization.

There are various techniques of meditation, and direct insight, which enable the realization of the pure formless immanence of existence largely devoid of individual forms. Through the mental exercise of meditation one greatly reduces the number of forms and more easily realizes the underlying field of pure immanent consciousness within which all forms arise. When forms disappear all that remains is the pure field of consciousness itself that is the experience of the pure field of formless existence within which all forms arise.

The experience of formless immanence as pure formless consciousness is essential for realization, but forms must be dealt with in daily life so it's also essential to realize the immanence manifested by individual forms. The realization of the immanence of both forms and formlessness is essential to the full realization of the immanence of existence and its experience as consciousness itself (Suzuki, 1956).

Every one of us already experiences the immanence of everything that exists in every moment of our lives as our consciousness of it. It's simply a matter of realizing what we are already experiencing. Consciousness itself is our direct experience of the immanence of existence in whatever form it takes. We realize the immanence of existence as consciousness in our consciousness of it.

Thus our consciousness itself is our direct experience and realization of the immanence of existence, both in all the individual forms of the world, and in the formlessness of an empty mind in meditation. Both are essential aspects of the realization of consciousness and immanence.

THE REALIZATION OF EXPERIENCE

To us the entire observable universe and every last thing in it without exception including ourselves exists only as our conscious experiences of it. All we are ever aware of is conscious experiences of information forms. Thus all that *observably* exists is consciousness and the information forms flowing through it.

No matter what the content of the information that flows through consciousness it's always just information. The information can be that of physical things, living organisms, theories of reality, or even a personal self having experiences, but ultimately it remains only the conscious experiences of the information of whatever it is.

In all cases all that ever appears in consciousness is information. Thus we never experience actual things, but always only experiences of what we take to be things. Experience and consciousness is always only of information representing things rather than things themselves. This is consistent with the notion that things are actually only the information of what they are, in this case the information of their representations in our simulation being experienced and even this also is ultimately only the experience of that idea.

Of course we can reasonably construct consistent theories of a world external to personal experience populated with other beings having their own experiences unknown to us but ultimately every one of those theories observably exists only as the conscious experiences of it. This is true of everything so in the last analysis all that confirmably exists is experience itself.

Thus experience itself is primal and fundamental, and the entirety of reality observably consists only of experience. This includes any notion of any reality that exists beyond experience because even those notions consist only of the experiences of them. The information content of experience can take any possible form, but whatever form it takes the fact of experience is fundamental. The information contents of consciousness continually change, but the consciousness they appear within remains the same.

Thus conscious experience itself is more fundamental than the contents of consciousness that appear within it. Consciousness itself is fundamental and exists prior to any of the information structures that appear within it. If consciousness didn't exist no information could be experienced. But even if the flow of information ceases consciousness still exists.

Thus all that observably exists is consciousness itself. The field of conscious experience is the fundamental substrate of observable reality; it's the medium in which all the information forms of experience are able to appear. Thus the field of consciousness itself is our direct experience of the substrate of existence within which all the information forms of existence appear. Consciousness is the direct experience of existence manifesting information forms of existence. Consciousness is the direct active self-manifestation of the immanence of existence.

The experience of consciousness is primal and fundamental and exists antecedent to any categories of information that appear within it including even the category of self. Thus we cannot even confirmably say that *we* are having these experiences because the whole notion of a self in a world of not-selves is just another information structure that appears as an experience within consciousness.

Thus experience itself is antecedent to any distinction of experiencer and experienced, self and not-self, individual things, or any distinction at all. From the perspective of experience all is consciousness only. All that can be observably confirmed to exist is ultimately conscious experience itself. All else is the information contents and structures that arise within consciousness but consciousness itself is always fundamental to them all. [That 'all is consciousness only' is also a tenet of the Buddhist doctrine of Yogācāra (Wikipedia, *Yogacara#Yogācāra in East Asia*).]

This can be easily realized in direct experience because it's what direct experience actually is. It's simply a matter of recognizing it for what it is at the fundamental level rather than being distracted by its current contents. At the primal level all that can observably confirmed to exist is conscious experience itself and the information flowing through it in the present moment. This is the realization of experience.

So all is experience but we can't convincingly call this experience *our* experience because it appears prior to any distinction of self and not-self. As Bishop Berkeley (Wikipedia, *Solipsism*) pointed out we can't even confirm with certainty that a world external to 'our' experience exists. He correctly pointed this out but the error of solipsism is assuming there is a self that's having experiences when the correct realization is that conscious experience itself is antecedent to the division into a self and not-self so all that confirmably exists is conscious experience itself prior to any distinction of self and not-self.

Even so we normally interpret conscious experience as the presence of a subjective sell having experiences though this is an optional add-on and not always present. From an objective standpoint all animals must have the conscious experience of a subjective self as a necessary component of computing appropriate actions as well as an internal model of an objective self though that may not be as explicitly conscious to the extent it is in humans.

Thus all that observably exists is experience itself and the information forms flowing through it in the present moment. Experience is the flow of information forms through the field of consciousness. And conscious experience is fundamental and prior to all distinctions and categorizations of information into self and not-self, experiencer and experienced, my experiences as opposed to other beings' experiences, and to any information structures whatsoever.

Conscious experience is all that observably exists and is simply the self-manifesting observable presence of the field of existence within which all information forms arise. Even so experience by its presence appears to manifest as a subjective self *having* experiences. So what then is this subjective self and what is its true nature?

THE REALIZATION OF CHI & ENERGY BODY

We are all familiar with our experiences of the internal feelings of the individual parts of our body. We can easily direct our attention to feel our arms, feet, hands, and other parts of our body from the inside.

Now if we combine the internal feelings of all parts of our body we can experience ourselves as a single *energy body*, which is simply the feeling of our entire body from the inside. This can be done in any situation with a little practice but may be experienced more fully lying peacefully on one's back without distraction. Our energy body is a simple straightforward experience we all have all the time if we just pay attention to it. There is nothing at all esoteric or metaphysical implied. Energy body isn't anything strange, new, or esoteric. It's simply what we always have been recognized for what we actually are.

Though we become aware of the internal feelings of individual parts of our bodies especially when we feel a problem, many of us seem resistant to taking the leap to experience all the internal feelings of our whole bodies together as a single energy body. But that is exactly what our own direct experience of our bodies actually is. The experience of the entire energy body from within is our direct experience of who we really are and is simply the experience of the active immanent presence of existence within us.

It's useful to have a term for the active immanence of existence within us and we may reasonably identify this with the Oriental concept of 'chi' if we are careful not to include any of the many irrational and exaggerated claims so often associated with it. In this usage chi is simply the energy of our active life force, and that of all living beings. It's simply the immanence of our existence, and every one of us experiences it all the time as the energy of the internal feelings of our bodies.

Thus chi is simply a useful term for the presence of existence within a living being. Chi is the same immanent existence that makes all things in the universe actively real and actual. We just feel it directly in our own selves as a distinct presence and call it chi.

While it's easily demonstrated that our experience of the flows of chi within our body are subject to some control by mind, breath and movement, one needs to take all the many claims about chi with a very big grain of salt and always subject to experimental confirmation. However there is good evidence that one can improve one's general health and well being by freeing the internal flows of chi through the energy body and by changing the tone of chi to feelings of love and well being flooding one's being rather than hostility, anger, hate, resentment, depression, or stress.

However such benefits are limited because all processes have their own chi, not just our own. All organisms including bacteria, viruses, attackers intent on harm, and even dangerous natural forces as well, all have their own chi energies, and so one's own chi is never a magically effective force against all harm. One needs to deal with the real actual energies of other processes and learn to avoid, redirect, or transform the harmful ones as best one can rather than assuming that just by strengthening one's own chi one will always prevail.

So our energy body and chi is our direct experience of our true self from within. From the perspective of direct experience 'we' are the total unified body of our internal feelings rather than the mental construct of our objective physical body that we usually identify with. Our energy body is what we actually are in our direct experience.

What our mind tells us is our 'physical' body is actually a mental construct it usually locates roughly in the same place as our energy body. As an aside, this easily explains how 'out of body' experiences occur. Our simulation normally collocates its construct of our objective physical body with our energy body, but since our mind does that arbitrarily as a matter of adaptive functionality, it's then easy to understand that in times of extreme immediate threat the simulation can just as easily relocate our consciousness out of our bodies in a attempt to lessen the potential experience of impending trauma.

In any case our energy body is not completely conterminous with how our mind represents the boundaries of our physical body but is generally felt as a field of feeling extending a little beyond the outlines of its representation of the physical body. There is plenty of anecdotal evidence that the energy bodies of other living beings can be felt beyond the boundaries of the physical body but little scientific confirmation.

Now in addition to the internal feelings of all parts of our bodies, all our feelings and perceptions of the 'external' world also actually take place within our own energy bodies. The touch of something our mind tells us is external is actually a feeling in our own energy body, and our entire sensory experience of the 'external' world actually consists entirely of feelings within our own energy body.

So our energy body is our experience of chi in all the internal feelings of all parts of our body. But it's also our experience of chi in all our sensory experiences of anything without exception. And the whole energy body also includes the experience of chi in every one of our thoughts, emotions, actions and all our mental states without exception. Thus our energy body is our complete whole being experienced from the inside. Our energy body is the totality of our experience because every bit of experience happens within us. We are our energy body and our energy body is what is having every one of our experiences. Our energy body is our feeling and experience of our total existence from the inside.

Though we may think of ourselves as our physical body in the fundamental reality of direct experience we are our energy body. We live within our energy body and we move within our energy body in everything that we do at every moment of our entire lives. We are our energy body consisting of conscious experience only and that is all we are at the fundamental level of direct experience.

Thus we are our energy body walking down the street, we are our energy body driving our car, we are our energy body meeting our friends, conducting our business or writing or reading these words, or relaxing on our couches watching TV. We are our energy body in everything we do in every moment of every day throughout our entire lives if we simply direct our attention to who we really are in our direct experience. This is the realization of energy body.

Thus every experience that occurs is the conscious experience of some form of chi, of the immanent existence of the information of what it is. And every experience is another manifestation of chi energy because chi is just a name for the immanence of existence as it's directly experienced.

Thus to realize our true self, our true nature, we simply become our total energy body including all its experiences. This is the realization of a self consisting of conscious experience only and including every last experience that observably exists. It's the natural manifestation of all conscious experiences as a self. If we are anything at all this is our true fundamental nature.

To complete the realization we must drop the 'our'. Conscious experience is antecedent to any distinction of self or not-self, or experienced and experiencer. Thus the energy body is true nature, but not 'our' true nature. It's true self, but not 'our' true self. The energy body is true self or true nature period without any specification of what. It's simply all that observably exists.

Thus if we seek a true self it's ultimately to be found in the total abandonment of a personal self. This is the mind of Buddha existing as pure conscious experience in a formless world through which forms pass manifesting the immanence of existence (Suzuki, 1956).

This realization is what the Diamond Sutra calls 'Awakening the mind while dwelling nowhere' (Suzuki, 1956). There is no central locus to consciousness because consciousness simultaneously pervades the entirety of experience. Forms arise and fade in the field of consciousness but consciousness remains. There is a total openness to everything as the forms of experience flow freely through the field of conscious existence. Outside this there is not even nothing (Wilhelm, 1931).

So to realize true nature just reduce everything that exists to its pure raw primal conscious experience and discard all notion of an individual self. Conscious experience itself as it occurs in the present moment is true nature. This is consistent with the idea that all we ever experience of reality is our simulation of it and every last bit of it occurs within our own being.

So it's reasonable to assume there is a reality outside our simulation that can be realized through our simulation of it but even this exists only as the experiences of it. It's reasonable to assume there are other beings having experiences we don't experience, but even this exists only as the experiences of it. Experience is primal and only subsequently is it organized by the simulation into the appearance of a universe within which we exist, but in the last analysis every bit of that structure exists only as the experiences of it.

But this is *the realization of the retinal sky* or horizon, that everything we seem to see in the world around us actually exists in our own simulation in our own heads, and is projected out around us to help us make better sense of the world.

Thus the entire world of appearances is an illusion. When we look out into the world we are actually looking into the interior of our own mind, into the interior of our simulation of reality rather than into reality itself. Go ahead and look. Every bit of what you see out there is its simulation in your brain. Bright moving meaningful colors painted over a highly simplified mapping of the actual logical structure of reality and projected out into an illusory 3-dimensional world that's neither within nor without.

When convenient to function more effectively we can fall back half way to the realization of our energy body as an individual self in a world of illusory appearances as our conscious experience of the totality of our internal feelings, thoughts and experiences sitting here in our chair or walking down the street or doing anything at all.

No longer are we a physical body viewed from the outside as an object. We are a living energy body experienced from within and moving from within through a world of experience only. And everywhere we look there is only experience and consciousness. We are our energy bodies in every instant of our existence and our energy body is all that observably exists. Our true nature is the true nature of the universe itself, the immanent experience of existence self-manifesting within us as it does within everything.

So move not as your physical body from the outside but become your energy body from the inside in the midst of all experiences. Walk down the street as your total unified flows of chi moving your energy body from within. Walk down the street as your Buddha. This is the realization of energy body and chi. This is our Buddha Nature.

BUDDHA NATURE

Having discussed realization from an objective viewpoint it's also useful to consider how some concepts of personal myth may aid in realizing it. Personal myths can assist in a more personal directed relationship with reality, and they can be quite useful so long as they are understood as myth rather than objective truth.

The concept of Buddha Nature is simply an ancient recognition of energy body. So there is a natural way to integrate this core concept of Buddhist tradition into Universal Reality. Buddha-nature can be easily identified with the immanence of existence, the universal active ingredient common to all things. But again the usual overlay of religious dogma and superstition that runs through most Buddhist sects must be carefully excluded.

Buddha-nature is a concept that often draws legitimate scorn among Western thinkers due to its many unscientific and illogical interpretations (Wikipedia, *Buddha-nature*). But again when defined rationally in terms of established concepts, Buddha-nature can be a useful aid in understanding and promoting realization because it enables a personal directed perspective on abstract concepts such as existence and immanence.

In our usage Buddha-nature is simply another name for chi, energy body, or existence from a more personal and individual perspective suggesting the possibility of personal realization and improvement. Thus the realization of Buddha-nature is another term for the realization of the true nature of things including one's self.

Though most Buddhist schools use a more restrictive definition, limiting Buddha-nature only to sentient beings, by our definition all things have chi or Buddha-nature because all things are forms that have immanent existence. This definition enables a simpler and more consistent view of reality, as it's just another perspective on what has already been established.

204

From this perspective realization can be considered the direct awareness or experience of the Buddha-nature of all things as the true fundamental actuality that fills the emptiness of their forms with the reality of being. This is consistent with the views of the more rational and philosophical forms of Buddhism such as Zen (Suzuki, 1956).

In this view all the things and beings of the world share the same presence of immanent existence as their common fundamental nature, and realization is the realization and experience of precisely this. All things share the same existence and this is true no matter whether their forms interact in harmony or in conflict with one's own form.

Thus realization involves seeing the Buddha-nature in all things and beings no matter who or what they are. As another name for reality itself the Buddha lives within the forms of all beings. Buddha bum, Buddha whore, Buddha killer, Buddha next door. All forms are manifestations of Buddha because all forms have Buddha-nature because the common fundamental nature of all things is existence and when we realize this we experience all things and beings as Buddhas whether they know they are or not, whether they have attained this realization or not.

This includes all animals and other organisms as well as people. Buddha bear, Buddha fox, Buddha bird, Buddha dog, Buddha cow, Buddha worm, Buddha flower, Buddha bacteria. Buddha nature is trapped within all forms and dwells in all beings waiting to be awakened to its true nature. And Buddha-nature is the true nature of every non-living thing as well. Every stone, every drop of water and speck of dust is a form filled with the Buddha-nature of immanent existence. In this view the entire universe of forms consists only of myriad forms of Buddha.

From this perspective we also have Buddha-nature and are Buddha. Buddha lives within us all and we can consciously choose to realize and express our Buddha-nature in a clearer, purer more realized form. We can abandon the unnecessary unhealthy forms that burden our personal self and become our Buddhas and move through every aspect of our lives as Buddha. We can be Buddha walking down the street, recognized or completely unrecognized through the world of forms. We can choose to let our Buddha guide our actions as we go about our daily lives as Buddha.

By realizing and becoming our Buddha, we are simply identifying with our higher truer selves and Buddha guides our actions from within. By surrendering our personal desires, attachments and prejudices to our Buddha-nature we become our Buddha and let our Buddha guide our actions, our lives, our work, and our destiny. We walk down the road as an empty form filled with the living immanence of Buddha being. In any case we are doing that already whether realized or not. It's just a matter of realizing it.

Of course this is all personal myth, a personal perspective on reality, and though certainly a useful aid to realization, we must be careful not to stray too far into fantasy. After all the Buddha within things can express only through the actual forms of those things. There are no super-heroes here. But there is nothing wrong with personal myth so long as it's recognized for what it is and doesn't lead us into delusional thinking but is used to inform and enhance realization.

With that caveat in mind then by becoming the Buddha we already are we become our true realized being in the disguise of our old self moving through the world of forms among other Buddha beings most of whom are ignorant of their Buddha-nature.

Our old personal self was an illusion of internal mental forms programmed into us since childhood. By becoming our Buddha our personal forms are transformed and purified by the flows of purer less mediated chi energy that naturally tends to manifest as a loving healthy life force. We swim like fish through the surrounding sea of living immanence, warm, loving and supporting. As our Buddha we realize we are empty forms within a warm loving sea of chi which continuously fills us with reality and being and we become better able to release and dissolve away all our stagnant unhealthy personal forms and blockages to allow chi to flow more smoothly and strongly and peacefully through us helping keep us vital, fresh and healthy. In this way, as our Buddha, our forms become purer, more balanced and strong.

By becoming our Buddha and living as our Buddha-nature we discard the illusory shells of our old personal being that concealed it from us. We see the world as it is with Buddha's eyes, touch it with Buddha's fingers, and manifest Buddha's realization of his own Buddha-nature as our true selves in everything we do. In this way we commune directly with the fundamental nature of reality itself as it self-manifests within us as our Buddha-nature.

DEFINING GOD

In Universal Reality there is no necessity of a God. The universe works quite well on its own, and certainly needs no external supernatural agency to design or operate it, nor does it need a creator since existence has 'always' existed. However the notion of God has a wide traditional appeal and for those in the monotheistic tradition there is a simple, reasonable and scientific way to integrate God into the theory if desired.

All that needs to be done is identify God with the universe itself, with existence, and with the motive force of happening of the universe. We then have a God which creates the universe of forms, is the source of the laws governing the evolution of those forms, and which sustains, directs and generates its evolution. This God is also the immanent living essence of all things that gives them being. There are obvious similarities to the gnostic and mystical traditions of the Abrahamic religions (Wikipedia, *Gnosticism*) (Wikipedia, *Mysticism*).

By this definition God even maintains it traditional attributes. God is certainly omnipotent as the happening of existence is the source of everything that happens. It's omnipresent as it's present in every detail of the entire universe, and in a sense it's omniscient as knowledge consists of information, and this God of the quantum vacuum is the source of all information. In fact since it consists entirely of information, the universe can be thought of as the knowledge of itself, the knowing of itself, as the running program of the mind of God that continually creates the current information state of the entire observable universe essentially by 'thinking' it into being.

And if anything is divine and miraculous, it's certainly the universe itself and the immanent existence that animates it. The universe itself is certainly the proper subject of our awe and reverence and devotion. And the existence of the universe as it is including our own personal existence is certainly the ultimate miracle.

However it's critically important not to include the huge burden of non-scientific mythology that clutters the Abrahamic traditions. From a scientific perspective it's clear that traditional religions were simply humanity's original attempts at developing a Theory of Everything. They were the best prescientific explanations of reality the ancients could come up with.

Unfortunately religions have now become delusional belief systems by persisting though history in the face of a vast body of contradictory scientific evidence. Belief in religions most likely persisted because they were highly effective means by which rulers legitimatized control over their subjects as the chosen representatives of God on earth enabling them to rule by divine right. But believing in religions as a matter of faith in this day and age is delusional and dangerous.

Thus if we want a God, reality itself is the only reasonable and scientifically acceptable definition of God. It also has several very important and obvious advantages. First there can be no doubt that God exists since it's self evident that the universe exists. And second the attributes of God now become merely a matter of scientific discovery. Third, this definition of God is non-sectarian and non-divisive, and should be equally acceptable to anyone with an open science oriented mind.

Most of the interminable and often violent arguments over whether or not God exists, and if 'he' exists what 'his' nature is are immediately resolved using this definition, and the way forward is clear to determine the rest through the application of logic and scientific method.

So, though it's not a necessary part of Universal Reality, the identification of God with the universe itself can lead naturally to a more personal and spiritual relationship with reality and aid in our appreciation of its awesome wonder. We may obtain a more personal relationship with the universe by identifying it with God. From this perspective God is reality itself and the active happening that animates all things and gives them existence according to their forms.

From this perspective we can also realize our own true nature as that of God. If God is the immanence of existence then God lives and breathes within us all and only waits for our realization of its presence to appear. And God's divinity is our own true nature as well so that we can now truly say that God dwells within us, that we are God.

By this definition God personally manifests within all of us as life and consciousness and is the true self of all personal beings. By this definition we, and all things, participate in and manifest the immanent divinity of reality.

Defining God as the universe is just a way of conceptualizing and relating to reality in a more personal manner. As such it can be a useful form of personal myth. Personal myths can be useful aids and comforts and aren't inconsistent with reality so long as they are realized as myth and not confused with objective reality. Only when they are mistaken for reality do they become delusional and hamper realization. Otherwise, recognized as myth, they can be perfectly consistent with reality and even aid in its realization.

From this perspective God, being the existence within all things, looks at us through every eye and looks out through our eyes at the world as well. And God sees itself in every eye looking back at itself looking at itself in recognition of itself. In this way God recognizes and knows itself and the reality of the universe and we and all beings become the sense and knowledge organs of God that allow God the universe to experience and know itself consciously. We realize ourselves as the consciousness of God within us, as God is the active living essence of all things including ourselves.

This is true not just of looking and seeing but of the experiences of all our senses and our consciousness as well. All the organisms in the universe are the means through which the universe as God manifesting in those individual forms becomes able to experience and know itself and thus begins to become more self-aware. We, and all beings, are the individual distributed sense organs and minds of God through which God knows and experiences itself and God the universe gains self-awareness.

Properly understood there is nothing supernatural about this realization. God as the active existence of the universe exists in every form but is only expressed through the actual form of that form. God sees only out of forms with eyes and cannot see out of forms without eyes but since all forms experience other forms in their interactions with them, God experiences through all forms and is experience itself, but only in whatever form that experience happens in. This is entirely consistent with Universal Reality if we define God as the existence of the universe.

Because experience is the self-manifestation of reality, God can be said to create and self-manifest itself as the experiences of all things. This is the universe experiencing, and in some forms knowing itself, and this is how God manifests and knows itself and becomes self-aware.

Thus the universe and God is its own self-awareness of itself self-manifesting as experience. In a fundamental sense it's not even clear we can meaningfully speak of the existence of a universe of information forms absent its experience of itself because there is no way to confirm its existence or structure if it doesn't self-manifest and observe itself.

Thus reality is reality experiencing itself. And all of us and all organisms and all things and forms are part of this process of the self-realization of the universe and thus the self-realization of God as experience.

Thus God has both a non-personal ubiquitous formless aspect as the formless sea of existence, and innumerable personal manifestations as the immanent existence of the forms of all individual things and personal beings. We may sense the formless presence of God in the meadow but God remains unseen and formless other than in the actual forms in the meadow that God is manifesting as. We may realize the presence of God in the form of every being and thing and in the formless sea that supports them all. God's presence is felt in everything around us as the immanence of existence, but God never appears except as it manifests in the actual forms of the world as the immanence of their existence.

We may long for God to appear as a personal caring and protective being in full divinity with supernatural attributes but this never occurs because God manifests only in actual forms and all actual forms are natural and obey natural law. But we can take comfort that the actual reality of this universal God is enormously more profound than any traditional supernatural being.

This is an entirely rational view of God insofar as it goes but one must always be wary of the danger of imputing any of the traditional delusional supernatural characteristics of the Gods of traditional religions to this God. This God is more akin to the rational scientific God of Einstein, and is simply another name for the immanent self-manifestation of existence within which the universe of programs and information forms arises (Wikipedia, *Religious views of Albert Einstein*).

The complete fine-tuning of the universe is such as to allow realization of its true nature that it may itself realize its true nature and divinity through us. And that's equally true of all of us and of all life forms that exist or have existed to the limit of the capabilities of their forms. We are all bound together in the enormous web of universal experience and consciousness through which the universe knows itself. May that be an enlightened and compassionate experience!

The direct experience of reality itself as consciousness itself is the living presence of this God in a non-personalized form. It's waking into a world where the presence of God is tangible as immanence but being formless remains unseen. But then some person or even some animal opens its eyes and looks at us and God suddenly manifests in that personal form looking out through its eyes. And all the while we were looking in vain for God with our own eyes it was actually God that was looking through our eyes searching for itself! God manifests in both personal and impersonal form because there is nothing that isn't God and God manifests in all the forms in the world as the immanence of their existence, as the immanence of the existence of the information of what they are.

Thus God is the totality of all forms including all of us simultaneously acting as its innumerable sense organs and consciousnesses and the combined experiences of all forms of God of each other. It's all its forms experiencing itself and thus continuously self-manifesting its immanent formless nature to itself in the present moment of its universal presence.

In this way Universal Reality explains the gnostic and mystical experiences of the Christian tradition as the direct immediate experience of the immanence of the divine nature of the existence of all things (Wikipedia, *Gnosticism*). For the Christian mystics this often manifested as the direct experiences of the immanence of existence especially in their Christian symbols but saints like St. Francis of Assisi seem to have realized the immanence of animals as well. Thus the gnostic and mystical traditions are based in the fundamental nature of reality even though their individual symbolism and interpretations are often delusional (Wikipedia, *Mysticism*).

In this view the universe and everything in it is the living presence of God. There is nothing that isn't God. Every part of the universe is part of the miracle of God's existence and science is the study of God, of the miraculous nature of God, of the miracle of God's existence, which is the existence of the universe.

Sitting inside the quantum vacuum as it computes the observable universe within it, here if anywhere we glimpse the mind of God at work creating the universe on the fly in effect continually thinking it into existence. Here is the mind of the universe imagining the world into living existence in all its awesome beauty, majesty, and wisdom, and in all the wonderful divinity of its immanent reality.

We all interact with God all the time in our every action. God reveals everything in every event but we understand only a little of what God is revealing. Studying the workings of God the universe is a proper form of prayer. Look to the universe itself for knowledge of the workings of the divine.

From this personal perspective we can say that everything that exists is the manifestation and presence of God, and we also are the presence of god. God shines in the immanence of all things and looks at us through every eye, and we look back at the world through God's eyes, and god sees itself looking back at itself in every eye. God is the universe and we are God.

THE REALIZATION OF LOVE

Though chi is the single energy of our existence, it's experienced in many different tonalities as the information of how our bodies feel from moment to moment from the inside. Chi is an important and immediate diagnostic tool of the internal state of our being and all parts of our body. It's important for our well being that we pay attention to the feelings of our energy body and understand what it's telling us.

We can also exercise a considerable amount of control over how we experience our chi. Properly nurtured, chi can manifest within us as a wonderful feeling of health, well-being, and love throughout our whole energy body. We can experience our chi as the living presence of pure unconditional love within us. Not only is this the most wonderful feeling imaginable but also there is considerable evidence that it fosters our health and well being, though of course the effect of our own chi is always limited in the face of other active chi energies. The universe is all one computational flow of chi or existence in which our form, which manifests our personal chi, is but a miniscule part.

We can choose to experience the presence of chi within us as pure love and well-being, as a feeling that floods our being and refreshes and nourishes us. We can also imagine this as the presence of the living God within us or the awakening of our Buddha-nature so long as we remember this is personal myth rather than objective reality. In any case it's a wonderfully refreshing and transformative experience.

And objectively we can say that God, the universe, does love us simply because the universe is continually manifesting us into existence. This is certainly the ultimate act of love. We exist only as the unique result of vast uncountable and unknowable numbers of enormously improbable coincidences. One out of millions of sperm at each conception of every one of our billions of ancestors stretching back to the beginning of life, and the actual pair choices of each of the multitudes of possible ancestral matings, not to mention the uncountable myriads of quantum events back to the original complete fine-tuning of our universe; every one of these had to happen exactly as it did for us to be here right now in the present moment of our existence as we are. Our amazingly improbable existence in the present moment is the ultimate act of love, and can certainly be experienced as such. The universe, God if you wish, embraces us in the arms of existence and floods us with the pure unconditional love of the immanence of being.

With the proper understanding and caveats we are one with God and Buddha Nature as we already share their common existence. We are a part of the living God of the universe. From this perspective God is our essence and is within us at all times. There is nothing other than God within the universe, and every finest detail of our being is a part of God. If you don't experience this it's only because you don't let yourself experience it because God is the active experience of all things. From this perspective every thing that happens in the entire universe is an act of God, is divine and perfect and absolute, and every one of us continually lives within God as a part of God in the existence of the present moment.

This wonderful, beautiful, and enormously profound new vision of reality emerges naturally from Universal Reality as a personal relationship with the existence of the universe. Completely consistent with modern science, this vision incorporates all the pieces missing from the usual interpretations of science such as consciousness, the present moment, and the nature of existence to achieve a complete Theory of Everything that automatically includes the realization of the reality it reveals.

Thus God can be identified with the divine living essence in all things including ourselves. All we have to do is realize its presence and God appears within us and becomes us and we become God. God is always right here within us waiting to be realized.

In this view God is the existence that animates all things and shines its immanence within their forms. All things are empty forms filled with God. God breathes in our every breath, God moves in our every movement, God thinks our every thought, God feels our every feeling and our every feeling is of God. And God is love and can be experienced as love, as love that fills the empty form of our being.

THE REALIZATION OF ACCEPTANCE

Because reality is absolute in the sense that it's all that is or can be exactly as it is in the present moment it is always enough. This is always true no matter where we are or what our situation is. There is after all nothing else possible in the present moment than what actually exists.

When this is realized there is never a need for anything else or any sense of loss, incompleteness or anything lacking. Because reality is the very substance of our being it's all that is ever needed because it 's all we can ever have or be. The ever-present formless essence of reality is all that can be and thus when its true nature is realized its direct experience is all that one could ever want or need. It's our very essence and our only true self and there can never be anything else.

Forms come and go but the immanent essence of reality always remains and it's always enough. This is always true; even as one works in the world of forms to effect changes in those forms our own inner nature, our own true self, is always the immanent existence of reality, which never changes. Forms come and go but what can be called our Buddha Nature always remains and thus our true reality always remains. From the Western tradition God never leaves our presence.

Forms continuously arise, change and vanish into non-existence but the common immanent reality within which all forms exist is always present. Forms themselves are empty, transient and illusory. It's only existence itself in which all forms arise that is permanent and ever present and always available to us in our form if we just open ourselves to its realization.

Because reality is what is and absolutely so and cannot be otherwise than it is right now, realization accepts it as such as it must to be in accord with reality. The necessity and inescapability of absolute acceptance of what exists is an essential part of realization. This is true not just of the formless essence of reality but of the current state of all forms in the present moment. Once forms appear they absolutely are as they are and must be accepted exactly as they are if the true nature of reality is to be realized. Otherwise we deny the reality upon which we depend.

This need not keep us from working to effect change, it just means accepting that the forms we are working to change are the ones that actually exist. By accepting things exactly as they are we increase our capacity to change them.

Realization also includes the complete acceptance of ourselves as we are. We accept ourselves as we are by releasing our desires and attachments for things that are likely beyond our attainment. By releasing unreasonable desires and attachments and by accepting our situation in life as it is we release the forms that lead to suffering and come closer to realizing our true self as it actually is in the present moment. Our true self is the one thing that is always attainable and within our grasp if we just open ourselves to it and embrace it because it's what we actually are at every moment of our existence.

There is an ultimate bravery in the total acceptance of reality as it is and confronting its awesome absoluteness directly and completely. It's also the total acceptance of our complete and total aloneness in the eternal presence of God the universe. We are completely and totally alone in a personal sense because our personal forms are inherently distinct from all other personal forms, and yet we are always completely and absolutely one with the living presence of God the universe, because we share the essence of existence within our personal form in common with all other forms.

In absolute acceptance of what is we dwell in perfect peace in pure love in the present moment. In this state of completeness there is nothing more that is needed. Reality is always enough. It is always eternally fresh and real and alive and is always immediately available to us because it's already our fundamental essence. Even in a world of problems and troubles, illnesses and pains, the immanence of existence is always enough. After all in the ultimate analysis there is nothing else.

PURPOSE AND ETHICAL PRINCIPLES

The theory of Universal Reality naturally leads to some basic ethical principles and suggests a plausible, though speculative, purpose for our existence both as a species and as individual beings.

We are certainly sense and knowledge organs of the universe through which it becomes better able to experience and know itself. The universal program has evolved us and other sentient life forms and through us is able to become aware of itself. God, the universe is waking up with us, and it can reasonably be argued that this is the purpose of our existence. But if so, to fully fulfill this purpose our knowledge and experience of reality must be as complete and as realized as possible.

To this end it's natural that it is ethically 'good' to spread scientific knowledge and realization as widely as possible and diminish delusion and ignorance and suffering as much as possible, and to that end, to work to make the earth a sustainable healthy and protected environment to facilitate this.

By doing so we move towards a more and more self-aware and enlightened universe in which God, the universe has maximal awareness and knowledge of itself. We can also speculate that this is the purpose of the universe itself, to move from an originally unconscious state towards the eventual goal of a fully self-conscious universe, and that mankind is a step the universe has evolved in its progress towards this eventual goal. This is of course speculative, but it's certainly a reasonable hypothesis based on the evidence.

The original fine-tuning of the universe implicitly contains within it the seeds of this progression, as it's exquisitely fine tuned so its programmatic evolution naturally leads through innumerable coincidences of random choice to the emergence of intelligent life capable of knowing the universe that produced it. All the critical elements of this design lie implicit in the original virtual nature of the quantum vacuum that gave rise to the universal program of existence including us.

If our true destiny is to function as sense and knowledge organs of the universe then the more accurate and compassionate and enlightened we are the better is the universe's experience of its own reality. Each of us is a little fragmentary bit of God, a little bit of God's total mind and body, by which God knows itself and with realization becomes enlightened through us as we become simultaneously enlightened through the experience of God. Certainly this realization is its own reward.

There is no absolute good and evil in the computational universe. These are human concepts, which are always relative to some set of human standards. And it's often quite difficult to apply any set of standards because whether effects are good or evil is always a judgment by someone at some particular time and what is good for one is often bad for another. However there are generally accepted social norms from culture to culture that have evolved primarily to facilitate stable societies. These social standards are the primary references for good and evil around which individual standards tend to cluster.

The idea of karma, that good ultimately begets good and evil begets evil is not consistent with the actual laws by which information forms evolve. There may be some tendency in some cases for like to beget like but there are numerous exceptions and by whose standards are ethical results to be judged, and at what point in the continuously evolving network of events? There are innumerable examples of well-intentioned actions producing tragic unintended consequences. And there is certainly no reincarnation so there can be no karmic transmission from one lifetime to another.

Nevertheless it's possible to outline some general ethical principles in the context of realization. Certainly the first is to attain realization itself. While Zen correctly points out that enlightenment is not something to be attained, that is the view *from* enlightenment rather than from *the path towards it*. The corollary is the Bodhisattva ideal to promote realization among all beings and to minimize suffering. This can be done by example, by teaching, and hopefully by writing books like this one (Tsunemitsu, 1962).

Another very reasonable core ethical standard is protecting and fostering the sustainable health and viability of Earth's biosphere. This is arguably the single most important ethical principle in that it sustains and maximizes the health and existence of all known life. Earth's biosphere is the only known cradle of the convergent emergence bringing self-awareness to the universe. For this to flourish human society must become sustainably integrated with nature, and man must begin to tend the earth as a natural Garden of Eden and strive to develop a Heaven on Earth. It would be an enormous, perhaps irreversible, setback to the apparent direction of the evolution of the universe if humanity were allowed to destroy the viability of the earth itself with all that implies.

Another fundamental ethical principle is compassion, which tends to arise naturally from the realization of the common Buddha Nature we share with all beings This realization naturally motivates us to help alleviate the suffering of all sentient life forms including our own selves and to foster realization among them.

This principle of compassion has profound consequences for how one relates to other beings including the question of eating meat. One recognizes the living sentient spirit within all animals and their capacity for suffering but at the same time one recognizes that predation fills an essential natural ecological function. All individual organisms must die and that death both supports life by providing food for other life and also makes room for new life and creates the opportunity for better-adapted life. We must realize and accept the great plan of life and death as essential for the evolution of the universe, but we should always do so in a compassionate and intelligent manner that minimizes unnecessary suffering.

Difficult questions always arise and there are not always easy answers. If life itself can be considered the ultimate individual good is it better for an animal to have lived a good and happy life till it's humanely slaughtered for meat or is it better for that animal to have never had the joy of existing at all? And if animals are to be killed for meat is it better to kill thousands of small creatures such as shrimp or one large cow of equivalent nutritional weight? These are profound questions that should always be approached with compassionate empathy for the beings involved.

Good and evil are not simple and are always human valued momentary snapshots of isolated events in an enormous web of ever evolving forms. And these judgments are always relative to each other in complex interacting processes playing out over different time scales. In general realization and compassion for all beings including ourselves and for the sustainable environment of the Earth are the great universal goods our lives should attempt to foster.

Zen has a somewhat similar approach that individual purpose is simply acting in accordance with the underlying principles of reality and flows of existence (Watts, 1957). It is to act not so much from one's personal desires, attachments and programming but in concert with the greater programs driving the world of forms. In so doing one gives up much of one's personal agenda and acts as one's realized self, one's Buddha within. In this view our ultimate freedom consists in giving up our personal freedom to align with the greater flows of reality, and thus our own realization and service is an example to others helping to liberate them from suffering.

The traditional Buddhist notion of the Bodhisattva who upon realization returns to the world to spread realization by example is the prototype of this principle (Wikipedia, *Bodhisattva*). The notion is that by teaching, working with the poor and needy, or simply manifesting realization in the world one furthers realization and ultimately helps release sentient beings from suffering.

THE ENLIGHTENMENT EXPERIENCE

Because reality is completely absolute as it is and absolutely real and present, its direct realization often occurs with sudden profound intensity. It's been compared to the sudden shock of meeting a tiger on the road or suddenly looking into the eyes of God and seeing God looking back (Wu, 2005). An enlightenment experience is the sudden realization of the actual awesome presence of the absolute realness of reality in all its immanence.

What was previously understood only as an abstract concept is suddenly realized as the living here now presence of reality itself directly within and around one. The true nature of reality is directly experienced and not just intellectually understood. Zen calls this experience 'satori' but a somewhat similar experience is common to many religions (Suzuki, 1956). It can come as a sudden profound shock to consciousness as the veils of illusion suddenly drop away, the scales fall from one's eyes, and reality is suddenly revealed right here and now in all its awesome absolute realness as the living essence of everything.

Because reality is absolutely real and absolutely what it is the effective intensity of its realization is unlimited and dependent only on the capacity of the experiencer. Normally mind operates at a mundane level preoccupied with a continual procession of daily forms and tasks and doesn't allow consciousness to experience the truly awesome intense absolute realness of reality that is possible. Allowing consciousness to experience something of the true intensity of reality is normally reserved for sudden emergencies or extreme sports where maximum attention and engagement are required for personal survival. But these situations mobilize intense adrenaline rushes in mind and body that can't be sustained.

The enlightenment experience is superficially similar in its intense clarity of mind but rather than extreme fight or flight adrenaline surges there is instead a strong, clear, healthy relaxed readiness of life energy that vitalizes rather than drains. This is a state of balance and refreshment rather than a sudden dissipation of energy. One is continuously aware of the awesome absolute presence of reality but there is a complete and total ease and acceptance and a perfect easy equilibrium in resting within it as if one had finally found one's true home (Suzuki, 1956).

By letting go of the natural tendency of mind to damp down the intensity of the experience of reality one naturally experiences that intensity to the level of one's capacity. To achieve that one must open oneself completely to the presence of reality and embrace it. Though sometimes frightening this becomes much easier when we realize there is simply no alternative to existing within reality as it actually is no matter how we might attempt to escape it by distracting or dulling our minds.

Mind normally makes us wary of reality and the dangers it may hold but while it's certainly true there are many programs running in reality that can pose significant dangers to our individual existence, the actual presence of reality itself is completely benign and in fact embracing it more fully and intensely enables us to detect and deter hostile forms more effectively (Saotome, 1989). Thus a major impediment to the intense realization of reality is the fear of hostile forces within it and the illusion that if mind somehow damps the intensity of our experience of reality that somehow protects us from those dangers when the opposite is actually true.

Thus completely opening oneself and embracing reality and releasing the illusory fear of its presence is essential to its realization and simultaneously allows us to live more effectively within it.

This is the mind of the samurai, which abandons individual self and accepts the total and absolute presence of reality, including even the ever-present possibility of personal death, and in so doing is able to exist at ease in the present moment with maximum effectiveness. The ultimate bravery in abandoning the forms of self that seek to insulate it from reality attains maximum realization of reality and maximum effectiveness within it (Musashi, 1974).

ZEN MIND

Realization is not to be found only within the gates of a temple or the teachings of some sect or master. Realization is the direct experience of reality and thus may be found anywhere and everywhere in any thing at any moment. Reality is everywhere and all one has to do is look with realized eyes to see it. No technique or path or teaching is intrinsically better than any other or even necessary. Sitting in meditation can be useful but realization is not found just by sitting. Realization is to be found anywhere in every moment of our existence because realization is the experience of reality and everything that exists, including ourselves, is the presence of reality (Suzuki, 1956).

Thus there is no transmission of realization or enlightenment. Teachers can be useful in demonstrating and guiding one along a path towards realization but they cannot transmit realization itself. There is nothing to transmit when everything is already present. Reality continuously self-manifests itself and reality itself is the only true master. All one needs to do is open oneself to the here-now presence of reality and see it for what it actually is.

Realization is not only to be found through a master's kōan. Reality itself is the ultimate kōan in whose solution is found realization. The quantum kōan and many others are the subject of Universal Reality. Reality is the only master and it presents itself to us in kōans every moment of our existence. Reality is the ultimate unanswerable question, the ultimate unsolvable kōan, in whose disappearance lies realization. The solution is not in the answer but in the vanishing of the question; in the realization of the presence of reality as it actually is. Realization of the living presence of reality itself unmediated by illusion is the only possible answer. The answer lies not in words, though words can be a guide, but in direct experience (Legge, 2010).

This is the meaning of the Japanese Zen term, 'Mumon', which can be variously translated as 'no gate', 'the gateless gate', or 'the gate to emptiness' (Blythe, 1966). 'Mu' does not mean nothingness in the usual western sense, but refers to the emptiness of forms in which is found the true presence of being. Mumon means there is no gate that must be passed through to achieve enlightenment. And it specifically implies it's unnecessary to pass through the gated entrance of any Zen temple or monastery to achieve realization. Wherever we are we are already within the true reality we seek and all that's necessary is to open ourselves to it.

YOU ARE ALREADY ENLIGHTENED

This book is a comprehensive and detailed search for the true nature of reality. We have discovered that the apparent reality of the world we seem to exist within is an illusion created by our minds, and not at all like the actual world of running programs computing data within a formless sea of immanent existence. And in this last chapter we have pierced the veils of illusion of our simulation and discovered how to directly experience the reality hidden within them.

221

But there is one final secret to be revealed. Ultimately we must realize that our illusory simulation of reality is in fact our only actual direct experience of the true nature of reality. Yes our simulation of reality is an illusion, but that illusion, like everything in the observable universe, consists of real information structures filled with existence existing in the medium of existence. Thus our illusory simulation is as much a part of the reality of the universe as any other information structure within it. Finally we realize the most important lesson of all, that *illusion mistaken for reality is illusion, but illusion realized as illusion is reality.*

Like everything that exists our simulation is a real actual part of reality. So our experience of our simulation is our direct experience of reality as it actually is insofar as it can be experienced. It's the true nature of the only part of reality we have direct unmediated access to. Our simulation of reality is our direct experience of a real part of actual reality and it's the only aspect of reality we can directly experience.

Thus we do directly experience the true nature of reality in every detail of every experience. We just need to understand what it is we are actually experiencing. Thus we directly experience the true nature of reality in every last aspect of our experience. It's just a matter of realizing that and working to refine and improve our simulation to better represent the true fundamental nature of reality. That's exactly what Universal Reality attempts to do.

Our mind's simulation of reality is a magician's trick. The trick is absolutely real, but its reality is not as it seems. Likewise our illusory simulation of the world is absolutely real and is our only possible experience of reality. Thus realization is not a matter of trying to escape or deny our illusory simulation, it's a matter of understanding and experiencing its true nature. We need look no further than where we are already, but we must look with enlightened eyes.

This is the meaning of the Zen saying by Ch'uan Teng Lu, "Mountains are mountains again" (Suzuki, 1956). "Before I had studied Zen for thirty years, I saw mountains as mountains, and waters as waters. When I arrived at a more intimate knowledge, I came to the point where I saw that mountains are not mountains, and waters are not waters. But now that I have got its very substance I am at rest. For it's just that I see mountains once again as mountains, and waters once again as waters." (Watts, 1957, p. 126).

Originally we thought of mountains as the physical mountains our simulation told us they were. But then we discovered that mountains were actually information structures computed by programs running in the quantum vacuum. But now we finally realize that the mountains of our simulation are in fact what mountains really are. Our illusory representation of a mountain is the real mountain of our direct experience, but now we understand and experience its true immanent nature as well as its illusory appearance, and in experiencing the truth the world becomes much richer and much more real.

We are the dynamic information structure of our total program running in the immanent existence of reality, and our simulation of reality is an integral part of that program. Though all aspects of our program interact computationally with both internal and external programs of the world at all levels of our biological hierarchy, our simulation is our overall model and our conscious experience of reality. We can improve the accuracy and realization of the illusory nature of our simulation, but the simulation, however we experience it, is the complete actual reality of our experience. As such our illusory simulation of reality is the reality we have always sought.

Thus we are all already enlightened. We are all already enlightened because we all live in the actual reality of our simulation all the time, and always have. We just have to look around and realize that the true nature of the reality we seek is our illusory simulation of it seen for what it actually is. This is all that exists in our experience and everything that exists is by definition part of reality. We need only realize it for what it is. We are all living in the true nature of reality and directly experiencing it all the time and always have been. It's just a matter of realizing what it is we are actually experiencing.

Our simulation is the only part of reality we directly experience completely and accurately as it actually is. Its illusory nature is its true reality, and we already have the most absolute realization and direct experience of the reality of our simulation possible. We need only recognize it for what it actually is, rather than what it pretends to be. The illusion of our simulation realized for what it really is, is the reality we seek, and ultimately this is the only realization possible.

But of course this precludes nothing. Our simulation is dynamic and can be improved as our understanding increases, or as we transition from mundane life to meditation to realization. But no matter how the simulation changes, in whatever form it takes, it is always our ever present direct experience of the true nature of reality, because whatever it is, it's always the true reality of the present moment. The conscious experience of the information of our simulation and its illusory nature is the true nature of reality that's accessible to us.

So finally we realize that with realization nothing actually changes from how we saw it before. The universe is as it always was. We just now see the world around us with entirely new eyes, as the most profoundly beautiful and awesome presence imaginable. All things are now the living immanent information of what they are continuously interacting and evolving in concert to the music of a single Uni-Verse stretching back to the beginning of time mysteriously revealed in the vast computational information nexus that we are part of.

Finally we understand that in our search for realization of the true nature of reality, that we all continuously live only within reality and are entirely composed of reality ourselves. There is nothing, and can be nothing, that is not already part of the true nature of reality. Therefore we are all already enlightened and could not be otherwise. We have always lived within enlightenment. The realization we have sought has been with us all along. It's just a matter of realizing it and embracing it.

So there isn't any trick or effort to realization. We are all already enlightened. Everyone is already enlightened and always has been. Enlightenment is simply a matter of realizing we are already enlightened and always have been because there is nothing that is not the real and actual presence of reality lying completely clear and visible before us. Of course realization can be refined, but enlightenment is just experiencing reality as it actually is and it's always exactly as it appears.

Everything in the world, every experience is exactly what it is. Yes, it has a deep structure, and yes it carries hidden secrets and is full of illusions, which are also part of reality, but nevertheless the true nature of what we experience is exactly what we experience because reality is exactly what is in the present moment. Even if reality is illusion mistaken for reality, even *that is* the reality of the present moment. However the deeper realization is experiencing illusion as illusion and through this its deeper reality. That then becomes the experience of reality realized more clearly.

Realization is simply whatever experience exists in the present moment, as it is with or without any interpretation in the simulation because all interpretations are also only the direct experiences of themselves. And so on it goes. Direct experience includes even the direct experience of even irrational and mistaken cognitive interpretations as well, whether realized as such or not. Illusion taken for reality is illusion, but illusion seen as illusion is reality, but in all cases experience is reality.

Everything is illusion but everything is reality because reality consists entirely of illusion when it comes to forms. The empty illusory nature of forms is their reality, and their reality is the manifestation of the nameless immanent presence of reality in which all forms arise and manifest the true nature of the universe and all things in the universe including ourselves.

With insight, study and practice more and more of the true nature of our wonderful universe is realized but what we do experience right now exactly as we experience it, realized as such, that is the true reality of the present moment. Thus we are all already enlightened and it's just a matter of waking up and realizing we are already here in the presence of ultimate reality and always have been!

Ultimately all we ever experience is the immanence of existence itself. In whatever form no matter how humble, in truth or illusion, or in relative formlessness, ultimately all that exists is the immanence of existence. And this is our true nature and the true nature of all reality.

Ultimately all is the mystery of existence, and what a wonderful mystery it is!

Welcome to Universal Reality!

EPILOGUE - TESTING THE THEORY

Ultimately the theory of Universal Reality may or may not turn out to be correct, either in full or in part. But the author is convinced it's reasonable, internally consistent, consistent with accepted science and that it presents a simple and elegant unified theory of all aspects of reality both scientific and those of direct experience. In this respect it's a true Theory of Everything.

Scientific theories are evaluated by

1. How well they explain observations.
2. Their consistency with the vast body of experimentally confirmed and internally consistent scientific theories.
3. Their own internal self-consistency.
4. The absence of falsifying evidence.
5. The scope of what they explain.
6. Their elegance, beauty, simplicity, and inherent reasonableness.

On all these measures the theory of Universal Reality earns high marks.

Every new theory must be subject to experimental tests to either confirm or falsify it. For example Einstein's seemingly outlandish relativity theory would never have been accepted had it not made testable predictions of the bending of starlight by the sun's gravitational field (Eddington, 1928). In this case the tests were straightforward, Einstein was confirmed, and relativity quickly became an accepted theory.

Other theories have not been so lucky. For example the theories of evolution and plate tectonics languished for years without simple conclusive tests until gradually the evidence became overwhelming. This is the usual case. Science progresses slowly and carefully but in the end it always progresses.

The theory of Universal Reality faces the same challenges. Since it's mostly a completely new interpretation of accepted science in the framework of a much broader Theory of Everything it's difficult to isolate clear tests that could either confirm or falsify it. However a few possibilities do come to mind and there are no doubt many others that are potentially subject to experimental and perhaps theoretical falsification.

The inability to falsify these tests would lend considerable credence to the theory.

1. Confirming a slightly positive curvature of space would tend to confirm that the observable universe takes the form of an extremely large hypersphere as Universal Reality predicts.
2. Universal Reality's theory of Dark matter as a spacetime curvature produced by the uneven Hubble expansion of space around the edges of galaxies and galaxy clusters should be confirmable by measurements of the strength and distribution of dark matter relative to the motion of the galactic masses that produced it.
3. It may be possible to test the theory that black holes are spheres of maximal intrinsic spatial velocity that don't contain singularities.
4. Any detection of dimensional drift or other relativistic anomalies especially over large expanses of space or time would tend to confirm Universal Reality's theory of a computationally based absolute spacetime background with respect to which rotation and world lines are relative. And confirmation of the necessity of an absolute dimensional background in relativity itself will also be a confirmation of Universal Reality.
5. Confirmation of coherence among the wavefunctions of particles emitted by events would lend credence to the theory of how quantum events create spacetime.
6. A mathematical test of the theory's core notion that dimensional spacetime can be computationally created along with mass-energy structures by quantum events and is key to unifying quantum theory and general relativity would be strong confirmation. To those interested in unifying relativity and quantum theory I strongly suggest this is the correct approach.
7. Confirmation of the METc Principle that all forms of mass and energy can be consistently modeled as different forms of spatial velocity. In particular that mass fields can be accurately modeled as minute spacetime vibrations, and electromagnetic fields as minute helical spacetime distortions.
8. The ability to program a convincing computer simulation of all aspects of reality based on the systems design presented by Universal Reality. To the extent such a simulation can correctly model all the major aspects of reality would be the best test of all and a very strong confirmation of the theory. We have already taken some initial steps in this direction that seem quite promising.

However since much of Universal Reality is a new *interpretation* of established science and other aspects of reality it may not always be subject to experimental tests. However it can be tested with respect to its logical consistency with accepted science. Overall consistency over maximum scope across all aspects of reality is the true and ultimately only test of validity, as it is of all knowledge.

In particular the mathematical implications of all parts of the theory must be clearly stated and tested for consistency with the established experimental results of relativity, quantum theory and other relevant disciplines. Universal Reality is an extremely promising approach to a Theory of Everything that seems *logically consistent* with established science but its mathematical consistency must be confirmed as well.

There are other useful tests as well such as elegance, simplicity and beauty. Universal Reality is founded on a set of simple principles of universal scope, and scores high on these criteria. And it proposes a computational model that is elegant and parsimonious compared with many of the currently fashionable interpretations of reality.

Not only does it explain the universe of science quite well but it does do in a manner that intuitively integrates existence, consciousness, and the present moment, the fundamental experiential constituents of reality about which current science has had nothing meaningful to say.

It's only by continually testing our theories that we can ensure their validity and improve our knowledge to progressively converge on the truth. However one must always be careful to test against actual theories rather than their interpretations. These tests may involve trying to find reasons why Universal Reality can't possibly work, and those will be useful in bringing to light points that need further development, but I suggest the more fruitful approach with any new theory is to try to find ways to make it work.

The theory of Universal Reality appears to have much promise in that it explains so much so well from a single universal approach. The general approach has been to accept experimentally confirmed science but develop a completely novel unifying interpretation that incorporates all aspects of reality in a single Theory of Everything. The author believes Universal Reality is the best most comprehensive Theory of Everything available and urges everyone to put it to the test in every way possible and let me know the results. I also welcome all comments and questions and can be reached at Edgar@EdgarLOwen.com.

BIBLIOGRAPHY

Blythe, R.H. *Mumonkan*. Hokuseido Press, 1966.

Chomsky, Noam. *Aspects of the Theory of Syntax.* MIT Press, 1965.

Cornford, Francis. *Plato's Cosmology: The Timaeus of Plato*. Hackett, 1997.

Eddington, Arthur Stanley. *The Nature of the Physical World.* 1928.

Greene, Brian. *The Elegant Universe*. Norton, 1999.

Greene, Brian. *The Fabric of The Cosmos.* Vintage Books, 2005.

Halpern, Paul & Wesson, Paul. *Brave New Universe*. Joseph Henry, 2006.

Hawking, Stephen W. *A Brief History of Time.* Bantam Books. 1998.

Hofstadter, Douglas R. *Gödel, Escher, Bach*. Vintage, 1980.

Legge, James. *The Tao Te Ching of Lao Tzu (translation)*. Commodius Vicus, 2010.

Lovelock, James. *The Ages of Gaia*. Norton, 1995.

Misner, Charles W.; Thorne, Kip S.; Wheeler, Archibald. *Gravitation*. Freeman, 1973.

Musashi, Miyamoto, Harris, Victor, trans. *A Book of Five Rings.* Allison and Busby, 1974

Owen, Edgar L. *Spacetime and Consciousness*. EdgarLOwen.info. 2007.

Owen, Edgar L. *Mind and Reality.* EdgarLOwen.info. 2009.

Owen, Edgar L. *Reality.* Amazon.com. 2013.

Owen, Edgar L. *Relativity Made Easy.* Amazon.com 2013

Owen, Edgar L. *Understanding Time.* Amazon.com. 2016

Owen, Edgar L. *Universal Reality*, Amazon.com. 2016.

Owen, Edgar L. *Unifying Relativity & Quantum Theory*. Amazon.com. 2016.

Penrose, Roger. *The Road to Reality*. Knopf, 2005.

Piaget, Jean, *Logic and Psychology*. Manchester University Press. 1956.

Piaget, Jean. *The Child's Conception of The World*. Littlefield, Adams & Co., 1960.

Saotome, Mitsugi. *The Principles of Aikido.* Shambala, 1989.

Schroeder, Daniel V. Purcell Simplified, http://physics.weber.edu/schroeder/mrr/MRRtalk.html, 1999.

Susskind, Leonard. *The Cosmic Landscape.* Little Brown. 2006.

Suzuki, Daisetz. *Zen Buddhism*. Doubleday Anchor Books, 1956.

Thorne, Kip S. *Black Holes & Time Warps*. Norton, 1994.

Tsunemitsu, ed. *The Teachings of Buddha.* Mitutoyo, 1962.

Vilenkin, Alex. *Many Worlds in One*. Hill and Wang, 2006.

Watts, Alan. *The Way of Zen*. Pantheon, 1957.

Wigner, Eugene. *The Unreasonable Effectiveness of Mathematics in the Natural Sciences*. John Wiley, 1960.

Wikipedia contributors. *Wikipedia, the Free Encyclopedia*. http://wikipedia.org. (n.d.)

Wilhelm, Richard, Cary F. Baynes, trans. *The Secret of the Golden Flower*. Routledge & Kegan Paul, 1931.

Wu, John C.H. *The Golden Age of Zen*. Pentagon Press, 2005.

Edgar L. Owen was born April 1st, 1941 and quickly realized that reality is not as it appears. A child prodigy, he entered the University of Tulsa aged 15 and received a B.S. with honors in science and mathematics with a minor in philosophy at 18 before completing several more years of graduate study in physics and philosophy.

In the early 60's he moved to the Haight-Ashbury in San Francisco where he hung out with notables from the Beat Generation, and conducted an intense personal study of the nature of mind and consciousness. From there he traveled to Japan where he lived for three years studying Zen and Buddhist philosophy while subsisting as a ronin English teacher.

Upon returning to the US he began a career in computer science writing numerous programs in artificial intelligence, simulations, graphics, and cellular automata while designing and managing advanced computer systems for the New York Federal Reserve Bank and AT&T. He then left the corporate world to start his own software business marketing his own CAD programs, which he ran for a number of years. Currently he owns a premier Internet gallery of fine Ancient Art and Classical Numismatics at EdgarLOwen.com.

Deeply immersed in nature since childhood, and always considering it the ultimate source of his inspiration and knowledge of reality, he has served as Chairman of his local Environmental Commission and organized several campaigns to protect the local environment and its wildlife.

Over the last decade he has worked to combine and organize the results of a lifetime of study of the various aspects of reality into a single coherent Theory of Everything. He now spends most of his time writing and exploring the wonderful awesome mystery of reality and how it can be experienced more deeply and fully and enjoying his existence within it.

Edgar currently lives in Northern NJ in a big brick house on top of a hill where he communes with nature and enjoys the company of his wild visitors including the occasional human. Edgar is currently single and can be contacted at Edgar@EdgarLOwen.com.

www.ingramcontent.com/pod-product-compliance
Lightning Source LLC
Chambersburg PA
CBHW081112170526
45165CB00008B/2422